T0211812

INDIGENOUS KNOWLEDGE AND LEARNING IN ASIA/PACIFIC AND AFRICA

INDIGENOUS KNOWLEDGE AND LEARNING IN ASIA/PACIFIC AND AFRICA

PERSPECTIVES ON DEVELOPMENT, EDUCATION, AND CULTURE

Edited by

Dip Kapoor and Edward Shizha

palgrave
macmillan

INDIGENOUS KNOWLEDGE AND LEARNING IN ASIA/PACIFIC AND AFRICA
Copyright © Dip Kapoor and Edward Shizha, 2010.
Softcover reprint of the hardcover 1st edition 2010 978-0-230-62101-5

First published in 2010 by
PALGRAVE MACMILLAN® in the United States – a division of St. Martin's
Press LLC, 175 Fifth Avenue, New York, NY 10010.

Where this book is distributed in the UK, Europe and the rest of the world,
this is by Palgrave Macmillan, a division of Macmillan Publishers Limited,
registered in England, company number 785998, of Houndmills, Basingstoke,
Hampshire RG21 6XS.

Palgrave Macmillan is the global academic imprint of the above companies and
has companies and representatives throughout the world.

Palgrave® and Macmillan® are registered trademarks in the United States, the
United Kingdom, Europe and other countries.

ISBN 978-1-349-38311-5   ISBN 978-0-230-11181-3 (eBook)
DOI 10.1057/9780230111813

Library of Congress Cataloging-in-Publication Data

Indigenous knowledge and learning in Asia/Pacific and Africa: perspectives on
development, education, and culture/edited by Dip Kapoor, Edward Shizha.
        p. cm.
        Includes bibliographical references.

    1. Ethnoscience—Africa.   2. Ethnoscience—Asia.   3. Indigenous
peoples—Education—Africa.   4. Indigenous peoples—Education—Asia.
I. Kapoor, Dip.   II. Shizha, Edward.
        GN645.I525 2010
        001.089—dc22
                                                                    2009053914

A catalogue record of the book is available from the British Library.

Design by MPS Limited, A Macmillan Company

First edition: August 2010

10 9 8 7 6 5 4 3 2 1

Transferred to Digital Printing in 2011

CONTENTS

LIST OF FIGURES AND BOXES

FIGURES

BOXES

CHAPTER 1

INTRODUCTION

DIP KAPOOR AND EDWARD SHIZHA

ANY ATTEMPT TO WADE INTO INDIGENOUS WATERS in the Asia/Pacific and Africa is a daunting task on many levels—there are the demands of definition, of positionality/location/optics of the contributors/writing (the transgressions of *outsiders* and liberties taken), questions of boundaries and spaces, telling/not telling, knowledge divides/unity, essentialisms and cultural dynamisms, the politics of the analysis/depictions of relative power(lessness) and of colonization/resistance, and so on that this set of readings is perhaps no less/more successful at addressing or transcending as is probably the case with many of its predecessors. The collection seeks to be as much about the contributions of indigenous knowledges (IKs), struggles, and peoples today (in contrast to museumized-mothball approaches that negate contemporary critical colonial relevance and presence/gravity of the indigenous—see Grande, 2004; *Red Pedagogy*) as it is about engaging indigenous postmortems/critical appreciations and expositions of the colonial project of modernity and its tentacular manifestations; that is, amplifying the political genius that stems from multiple histories of addressing "colonial assimilation projects, neglect, diminishment, and racism" (Battiste, 2008, p. 85). There is an attempt here to *learn from* and *with*, to *politicize* and *amplify* rather than to *anthropologize* and to *stifle/objectify* in yet another act of colonial representation, that is, rather than settle for mere descriptive ethnography with no apparent purpose but to *mine* or engage in *academic voyeurism*, contributors have been encouraged to consider, through praxiological engagements, the *political and cultural projects of indigeneity* and the simultaneous *re/shaping and disembedding of modern* epistemic, knowledge, learning, education, culture, political economy, and human-ecology relations (to consider the *contexts of coloniality* see Mignolo, 2003; Quijano, 2000): how far the collective contribution and each contributor has succeeded in this task is a matter for the reader to ascertain, intentions and commitments notwithstanding.

While the academic works and political projects of indigenous peoples of the Pacific, including the seminal contributions of the Smiths, Linda, and Graham, and of settler-colony indigeneities in North America and Australia have *arrived* and continue to make their mark on the colonial project of modernity, including the decolonization and recoding of resistance (see Aziz Choudry in this collection, Chapter 4), there is a lot to be learned from the Asian and African contexts of indigeneity and the relatively unheralded and ongoing historical contributions of the post/colony, as is pointed out by Kaushik Ghosh in this collection (Chapter 3). It is with this in mind that the idea for the collection emerged. Relying primarily on diasporic and indigenous scholars from Asia and Africa and their relationships and associations with communities/peoples of *the* Kondh/Adivasi, the Chakma and Marma, the Dayak and the Moi, the Truku, African-indigeneity in Zimbabwe and the Sub-Sahara, the Newars, Dalitbahujans, the Bauls, the Haya and the Fulani and the herbalists from the Luo, Abaluhyia, Abagusii, Akamba, Agikuyu, Aembu, Ameru, Ambere and Mijikenda and the bonesetters of Northern Ghana, this collection represents a collective effort to magnify the contributions of indigenous peoples of Asia/Pacific and Africa.

We recognize that definitions of indigeneity are contested, in some cases ambiguous and in others, a figment and tool of the state's governing apparatus. However, self-definition and the struggles and aspirations of the groups who appeal to history/origins, priority, ethnicity, culture, culturally embedded political economies, common struggles with modernist hegemony, and its sentinels are *significant* and *present*—self identification is probably at the heart of indigenous identity and is the dominant approach utilized by contributors to this collection. Many would prefer not to define indigenous peoples at all because, after all, definition/taxonomies (and their rigidities) are a product/instrument of colonial administration and control, as *outsiders* continue the offensive practice of *bounding* the *indigenous* while identities themselves tend to be more fluid. In fact, the UN Working Group on the Declaration on the Rights of Indigenous Peoples expressed concerns and stated that a "definition of indigenous peoples was unnecessary and that to deny indigenous peoples the right to define themselves was to delimit their right to self-determination" (cited in McNeish & Eversole, p. 5). The notion of *peoples* underscores the great diversity (identities) signified by this term and the varied experiences (histories) of indigeneity across the globe, some of which may include (as perhaps being common to most indigenous peoples): being original inhabitants of a land later colonized by others; nondominant sectors of society with unique ethnic identities and cultures; strong ties to land and territory; experiences or threats of dispossession from ancestral territory; the experience of being subjected to culturally foreign governance and institutional structures; and the threat of assimilation and loss of identity vis-à-vis a dominant society (McNeish & Eversole); that is, indigenous understood as a *location* and a *historical and contemporary* experience with *colonialism as the category of common experience.*

While it is estimated that about 70 percent of the world's indigenous peoples are in Asia (IFAD, 2000/2001), defining indigenous peoples in this part of the world is particularly problematic (Barnes, Gray, & Kingsbury, 1995) as: (i) governments in the Asian post/colonies do not recognize the category (with the exception of the Philippines); (ii) Asia has experienced different waves of migration and multiple colonizations whereby one ethnic group may have longer-standing claims without actually being the original inhabitants of an area; thus indigenous here often means *prior* rather than *original* peoples; (iii) dissident ethnic groups sandwiched between indigenous and nonindigenous peoples obscure such claims by demanding indigenous status as well; and (iv) unlike in settler colonies (North America, Australia, and New Zealand) where the *indigenous peoples category* came into existence in relation to the relatively recent legacy of Western European colonialism and have a defined, long-standing, and recognized status as such, the indigenous peoples of Asia do not have such clear *definition*. Nonetheless, the identifier does and continues to serve many groups in Asia who would refer to themselves as prior inhabitants vis-à-vis later arrivals (Barnes et al., 1995).

In Africa, the term *indigenous* is seldom used. Despite the experience with overseas colonizations, native Africans, many of whom predate the arrival of European colonizers, do not self-identify as indigenous. Other terms such as *tribes* or *ethnic* groups are used more frequently (Eversole, 2005). According to Sylvain (2002), "indigenous peoples in Africa tend to be defined narrowly as those specific peoples that are non-dominant (vis-à-vis other ethnic groups) and have close ties to ancestral lands, including land-based livelihoods (hunting-gathering-herding)" (pp. 1075–1076). However, these criteria are also contestable and inconsistent among themselves as, for instance, marginalized/nondominant need not go together with close ties to land or vice versa. Nonetheless, the term indigenous peoples in this sense is more specific to the *San*, the *Hadzabe, Maasai,* and *Tuareg* (mainly pastoralists and hunters and gatherers), for instance, who have been marginalized relative to agriculturalists in both colonial and postindependence eras (Eversole, 2005). Similarly, according to Eversole (2005), citing the World Indigenous Report of 2003, the indigenous peoples of Ethiopia, including the *Somali, Afar, Borena, Kereyu,* and *Nuer* who are 12 percent of the country's population and mainly pastoralists, "are subjected to the worst forms of political, economic and social marginalization, subjugation and inequality" and "development often involves confiscation of their grazing lands and forced sedentarization" (p. 33).

While indigeneity is a status usually accorded to remnant prior peoples living on their former lands in the margins of nation states (Tomaselli, 2007), in Africa, according to Edward Shizha (2008), all citizens of African origin who are African not through being offsprings of settlerism and colonialism are accepted as indigenous people. These are people with a long history of African ancestry and identify themselves one way or another with rural communities. African indigeneity is not the universalized cultural representations that have been exported to non-African societies, but a cultural life embedded in authentic spiritual, ecological, economic, political, and holistic life experiences. It is a way of life that is embodied in "existential authenticity that is required for indigenous people to engage in self-defining and self-sustainable development projects" (Shizha, 2008, p. 39).

IK, like the definition of indigenous itself, is also variously discussed; while Mi'kmaq scholar Marie Battiste is critical of the European demand for definition, she does provide a conception that may throw a wide net when it comes to understanding IK in other regions: "IK . . . is the expression of the vibrant relationship between the people, their ecosystems, and the other living beings and spirits that share their lands" (Battiste & Henderson, 2002, p. 42). According to others, "indigenous knowledge(s) differ(s) from conventional knowledge(s) because of an absence of colonial and imperial imposition" (Dei, Hall, & Rosenberg, 2000, p. 7). Whatever the conception, the authors in this collection strive to provide a sense of IK and learning that is context specific and generated within and by a particular trajectory of social and historical relations, including and most often, colonial relations. While the significance of IK has grown in relation to the discourse of sustainable development and health knowledge and the growing significance of the "indigenous category" in international law and regulation (Barnes, Gray, and Kingsbury, 1995), not all contributors here frame IK in terms of its usefulness to *modern development* and the resuscitation of this project (IK as having instrumental worth), as IK is seen *in relation to and within* indigenous onto-epistemic conceptions of the *good life* and as having intrinsic worth while being embedded in indigenous starting points and political projects.

When it comes to formal spaces of learning, according to Shizha (2006, 2008) the disempowerment of African indigeneity and knowledge constructs by colonialism and contemporary globalization has promoted the displacement and silencing of African knowledge systems and pedagogical practices in academic institutions. While colonization attempted to marginalize as well as disrupt the spiritual and cultural beliefs and traditional ways of life of African peoples (Wane, 2008) and bury indigenous ways (social, cultural, political, economic, spiritual, ecological, and integrated experiences), the colonial project is a complete failure as indigenous practices still exist in much of contemporary Africa. However, the effects of colonization on colonized minds (Said, 1994; wa Thiong'o, 1986) is also acknowledged. Africa needs educationalists who are capable of deconstructing essentialized forms of dominant Western knowledge constructs that are uniformized, while disregarding the multiple locations and intersections of knowledge production. The importance of IKs cannot be overlooked or devalued in this respect. The problems that Africa faces in terms of poverty and economic marginalization, hazardous health challenges, and education may be attributed to the lack of focused programs that recognize that IKs are the bedrock of the African psyche and the way of life. The well-being of Africans lies in creating and recreating conceptual possibilities that are indigenous, and possibilities that provide counterclaims to the illegitimation and political marginalization of what is indigenous and pragmatic to African societies.

This collection represents one of myriad and growing attempts to contribute to the process of decolonization with the help of indigenous categories. The book seeks to affirm the importance of claims to indigeneity in Asia/Pacific and African contexts and the associated claims to the uniqueness of IK conceptions, given the history of colonialism and its contemporary manifestations and associated encroachments into indigenous space. Some guiding questions that the contributors attempt to address to varying degrees include: What is indigeneity in the Asia/Pacific and African regions? What is IK in these contexts today? What are some of the "essentialist" dimensions of IK and learning today? How have and do these essentialist dimensions maintain a continuous trajectory in these various locations? What are some of the tensions/politics between indigenous configurations and modernizing/colonizing compulsions of today? How are indigenous peoples continuing to address these historical and contemporary colonizations (prospects and realities for social change)? In what geo-temporal arrangements are such politicized engagements evident (e.g., in formal education, health, social struggles/movements contesting development as compulsory modernization, etc.)? How and what is being learnt in these encounters?

THEMATIC ORGANIZATION OF THE BOOK: PREVIEWS AND DIRECTIONS

The collection is organized in relation to five broad thematic areas that are not intended as watertight containers but as key organizing foci, with considerable cross-pollinations for the reader to ascertain, pertaining to the rich canvas of indigenous lives in various Asia/Pacific and African contexts. Chapters 2 through 7 mostly consider indigenous struggles/activisms and critique, emphasizing the critical political knowledge and learning—*indigenous dissent capital*—from and within political and cultural struggles (Choudry & Kapoor, forthcoming) or indigenous *life projects* (Blaser, Feit, & McRae, 2004) with contemporary modern re/colonizations of indigenous space (spiritual, material, and cultural) or the *development* project (and the globalization of capitalism) (Bargh, 2007; McMichael, 2006; Rajagopal, 2004). These chapters dwell on insights that are mostly drawn from the research and praxis of the various contributors, most of whom have long standing relationships with the *Kondh* (India) (Chapter 2) and India's *Adivasis* (Chapter 3), the *Maori* (Pacific/Aotearoa/New Zealand) (Chapter 4), the *Chakma* and *Marma* (Bangladesh) (Chapter 5), the *Truku* (Western Pacific/Formosa) (Chapter 6), and the *Dayak* and the *Moi* (Indonesia) (Chapter 7).

In "Learning from *Adivasi* (original dweller) Political-Ecological Expositions of Development: Claims on Forests, Land, and Place in India" (Chapter 2), Dip Kapoor develops a critical exposition of development in India's forests/rural regions that is emergent from *Kondh Adivasi* political-ecological knowledge and learning gained through encounters between *Kondh ways* and the contemporary neoliberal development/globalization projects (Menon & Nigam, 2007). He suggests that "[t]he [*Kondh Adivasi*] claim on the state is to recognize *Adivasi ways* of living and being as a political right and not as an essentialized-inferiority in need of protection and welfare from a self-appointed guardian or parternalistic state." Given the failure/unwillingness of the state to recognize this, his chapter points to the growing number and significance of *Adivasi* (and wider subaltern coalitions) struggles/movements addressing forests/land/cultural space and their role in continuing to undermine the "colonial prospects of the neoliberal development, globalization, and compulsory modernization" agenda. In Chapter 3, "Indigenous Incitements," *Kaushik Ghosh* considers the politics of indigeneity in India, between colony/postcolony (e.g., India) and settler-colony (e.g., Australia) conceptions/indigenous politics and the lost possibilities when it comes to the "predominance of the imagination of the indigenous derived from settler-colonies" in relation to indigeneity scholarship and the transnational indigenous movement facilitated through the UN, while the "achievements of the indigenous throughout colonialism and in the postcolonial world" are yet to be fully recognized. Perhaps this is a case in point for the need for a book collection of this ilk, a collection that seeks to aim the spotlight on indigenous contributions in Asian, Pacific, and African contexts. *Ghosh* attempts to map an alternative "archive of indigeneity" whose radical potential "lies in such alternate archives or configurations, the very assembling of which necessarily demands a realignment of the coordinates of the episteme and ontology of the non-indigenous itself."

In Chapter 4, *Aziz Choudry*, drawing on his "close collaboration with Maori sovereignty activists in confronting capitalist globalization," adds another dimension to *Kaushik Ghosh's* observations regarding the dominance of settler-colony indigeneity (e.g., Maori) in transnational movement spaces, by suggesting that Maori movement-critiques (critical political knowledge of Maori) of neoliberalism and the colonial state (New Zealand), which prefigured the rise of what is called the "antiglobalization movement" or "global justice movement" (transnational movements), has largely been ignored by "white progressive economic nationalists" in global justice movements "who still advocate for a retooling of the state . . . in a way which ignores the fundamentally colonial nature of the nation state." In "Against the Flow: Maori Knowledge and Self-Determination Struggles Confront Neoliberal Globalization in Aotearoa/New Zealand," *Choudry* provides an overview of colonialism and neoliberalism in New Zealand, the context for Maori resistance, and pays particular attention to Maori movement critical knowledge production regarding connections between colonization and neoliberalism (also see Choudry, 2009; Grande & Ormiston, forthcoming, on this point) by focusing on struggles over IK/ intellectual property rights, tensions with non-Maori environmentalism and NGOs, and examples of cultural forms of Maori resistance.

In "Ethnic Minorities, Indigenous Knowledge, and Livelihoods: Struggle for Survival in Southeastern Bangladesh" (Chapter 5), *Bijoy Barua*, drawing upon his long association with *Chakma* and *Marma* communities in Bangladesh, develops a critique of the "Western development model" and a "scheme of industrialization, modernization, and commercialization" that is beginning to challenge local ethical and spiritual values. Using the case of the Kaptai dam, he discusses the local economic and cultural implications of such modern development schemes in relation to the lives of the *Chakma* and *Marmas* and the role of Western science, financing, non/formal education and institutions (e.g., NGO-led penetrations) in promoting such development. *Barua's* insights also include a critical examination of the considerable resistance to the dam and development displacement in the area—a process of resistance that saw the emergence of a *Chakma* hegemony, given their numerical supremacy and the rise of opportunistic *Chakma* elites who discredited the resistance. The peace accords of 1997 have simply led to another round of NGO proliferation and microcredit (small "c" capitalism) schemes and educational interventions, which continue to deny locally embedded economies, cultures, spirituality, and ethical values.

In Chapter 6, "Animals, Ghosts, and Ancestors: Traditional Knowledge of Truku Hunters on Formosa," *Scott Simon* draws upon his (2004–2007) anthropological research with the *Truku*, to describe how *Truku* knowledge about law, human-animal relations, and their ancestors contributes to the reproduction of language and values and makes a wider knowledge contribution that is relevant to Taiwan. This chapter also briefly examines indigeneity in Taiwan in relation to the government's decision to take over *Truku* territory through the development of the "Taroko National Park." *Simon* elaborates on *Truku* hunting practices and spiritual beliefs, while pointing out the significance of the movement of the *Truku* and other indigenous tribes, as "their knowledge reveals the expansion of the state system as a historical process of colonialism"—a state whose "written laws permit mining and criminalizes hunting in the very same mountains."

Ehsanul Haque, in "Development Enterprises and Encounters with the *Dayak* and *Moi* Communities in Indonesia" (Chapter 7), points out the centrality of forests to the livelihoods of the *Dayak* of Kalimantan and the *Moi* of West Papua. *Haque* explores *Dayak* and *Moi* encounters with the drivers of development enterprises including mining multinational corporations and international aid agencies and suggests that indigenous groups in Indonesia are "being dispossessed of their traditional lands . . . due to rampant logging, agro-industrial developments (palm oil production), and extensive mining carried out by trans/national companies." *Haque* also elaborates on the responses/resistance of the *Dayak* and *Moi* and how these struggles seek to address the rights and interests of indigenous people in Indonesia.

Formal education in schools and universities serve as other sites of epistemic and knowledge contestations and cultural possibility (Abu-Saad & Champagne, 2006; Villegas, Neugebauer, & Venegas, 2008). While struggles in relation to the development project (political economic and political ecological aspects or a *material politics*) draw upon integrated knowledges/learning embedded in indigenous *life projects* and community (or what in some academic quarters would be referred to as informal and incidental learning—ways of describing certain dimensions of IK/learning processes perhaps), *Edward Shizha* suggests (Chapter 8) that IK has been found to play a major role in cognitive development and contributions to successful and meaningful learning in both formal and informal settings. Everyday life experiences are rooted in individuals' social and cultural environments, which have a lot to do with how people interpret and construct meanings from new situations. Educational settings which take cognizance of the everyday life experiences of learners in terms of their oral traditions, indigenous/cultural knowledge and socioethical values, historical commemorations, and modes of media used to give voice to their inner-selves will go a long way in liberating educational and learning spaces. In "Rethinking and Reconstituting Indigenous Knowledge and Voices in the Academy in Zimbabwe," *Shizha* focuses on IK recovery in the academy as an anticolonial and antiracist project that needs to be pursued to liberate the minds of indigenous academics and researchers who have wittingly or unwittingly adapted Western knowledge as the universal knowledge and the panacea for underdevelopment in Zimbabwe. *Shizha* argues for integration of IK with academic knowledge in higher education in Zimbabwe to recapture the neglected historical contributions of indigenous Zimbabweans.

In Chapter 9, "Education, Economic Development and the Newari People of Nepal," *Deepa Shakya* echoes *Shizha's* argument on the marginalization of indigenous perspectives in formal education. *Shakya* argues that centralized education curriculum tends to undermine indigenous/cultural knowledge and socioethical values of indigenous communities in Nepal. This centralized education curriculum policy fostered dependency rather than sustainable development among the *Newars* in the name of economic growth and development. The author analyses contradictions in the *Newari* culture, modern education, and economic development and *Newari* community perspectives and extends the analysis to the contradictions of the process of socioeducational change across/between cultures.

Beyond the *formal dimensions/spaces* of *walled-systems of education* for credentialing or in Yup'ik scholar, Angayuqaq Oscar Kawagley's terms reiterated by Ray Barnhardt (2005) while formal education is "about making a living," IK and learning are also about "making a life" (pp. 113–114); that is, learning through various communicative mediums, including oral mediums, songs, and indigenous media, and through mediums that blend modern technologies into indigenous projects as through the incorporation of video technologies—these are the preoccupations of the next three chapters by *Ali Abdi* (Chapter 10), *Souryan Mookerjea* (Chapter 11), and *Sudhangshu Roy* and *Rayyan Hassan* (Chapter 12).

In "Clash of Oralities and Textualities: The Colonization of the Communicative Space of Sub-Saharan Africa", *Ali Abdi* builds on previous works (see for instance, Abdi in Abdi, Puplampu, & Dei, 2006) pertaining to processes of learning, enculturation, and *selective Africanizations* in educational projects of decolonization by underscoring the manner in which communicative space in sub-Saharan Africa (with a wider pertinence for other contexts of indigeneity) has been and continues to be colonized by the privileging of *text* and the simultaneous depriviledging of *oralities* and in the process, oral communities. In contrasting these two communicative realities, *Abdi's* chapter "advances the richness of oral traditions" and speaks to "how they have been trumped by the imposition of textual forces that, on the basis of selfish colonial interests, were bent on de-epistemologizing and by extension, de-ontologizing the historico-culturally located primary communicative lives of subordinated populations."

In Chapter 11, "Autonomy and Video Mediation: *Dalitbahujan* Women's Utopian Knowledge Production," *Sourayan Mookerjea* examines the reproduction of subaltern or IK through video-based media of a group of *Dalit* women farmers from Andhra Pradesh, India, who belong to a media cooperative called the Community Media Trust (CMT). *Mookerjea* provides some historical background that locates the women of the CMT and their primary audience, the marginalized farmers of the region, in relation to the persistence of class, caste, and gender inequalities of postcolonial India and with respect to their contemporary struggles against exploitation at the hands of globalized agribusiness and the neoliberalized Indian state. He focuses on the mediation of subaltern knowledge and examines the specific political-aesthetic features of the CMT's practice of *autonomous video* that enables subaltern knowledge to be conserved through its production and politicization. Similarly, in "Voicing Our Roots: A Critical Review of Indigenous Media and Knowledge in Bengal" (Chapter 12) *Sudhangshu Roy* and *Rayyan Hassan* examine the survival of indigenous media in Bengal within the sociocultural context of colonization and globalization. Specifically, the chapter investigates the survival and growth of *Jatra* (folk theater) and *Baul gaan* (songs of *Bauls*) as examples of indigenous media. *Roy* and *Hassan* conclude that indigenous media in Bengal has been providing social, philosophical, and spiritual education to the people of the region well before recorded history. With this in mind, the authors identify factors within the traditional Bengali media forms which ensure their survival in the current modern state and neoliberal cultural context, while also discussing the significance of indigenous media and the prospects for decolonization in the greater Bengal region.

Discussions on IK and its importance among indigenous people of Africa, Asia, or any other continent would be incomplete without considering gender. Gender is an important element in the social organization of indigenous/communities. The way knowledge is constructed and apportioned depends largely on the distribution and performance of gender roles. In Chapter 13, "*Haya* Women's Knowledge and Learning: Addressing Land Estrangement in Tanzania," *Christine Mhina* discusses the marginalization of women's roles in indigenous African communities, with particular emphasis on the *Haya* in Tanzania. *Mhina* argues that women in African communities are active participants in the economic production of the community. She illustrates the agency and ability of indigenous women to know, to choose, to imagine, to create, to decide, and to act. *Mhina* uses the Tanzanian situation to show local women's creativity and resourcefulness to develop indigenously informed collective solutions to address their marginalized rights to land tenure. The chapter highlights women's marginalized position within the existing *Haya* indigenous land tenure system and how colonial state impositions have further marginalized *Haya* women in relation to access and control over land. *Mhina* concludes with some reflections on the prospects for change with regard to addressing land discrimination against *Haya* women.

In Chapter 14, "The Indigenous Knowledge System (IKS) of Female Pastoral *Fulani* of Northern Nigeria," *Lantana Usman* examines the feminization of IKSs among the pastoral Fulani of Nigeria, and indeed across West Africa. Echoing *Mhina's* perspective on the agency of women in indigenous African communities, *Usman's* chapter discusses the cultural essentials that define female social agency and coexistence among the *Fulani*. *Usman* discusses and illustrates the purpose, nature, characteristics, and processes of IKSs in order to demonstrate the connection of learning and its possibilities for women and the community. The chapter also analyzes major IKS epistemologies (cognitive knowledge), axiology (value system), aesthetics (goods, beauty, and arts) and the idealism-realism construct (spirituality and the environment) practices of the pastoral *Fulani* women and girls. *Usman* concludes that feminization of the knowledge system is significant for the social and economic sustainability of women's lives and the promotion of their identities. She recaptures *Mhina's* theme on the undervaluing of women farmers, even though they play a significant role in social and economic organization in Africa.

Traditional medicine (health knowledge and learning) remains an important primary line of health service in indigenous communities throughout the world. Traditional medicine can be defined as the knowledge, skills, and practices of holistic health care, recognized and accepted for its role in the maintenance of health and the treatment of diseases, and it is based on indigenous theories, beliefs, and experiences that are handed down from generation to generation (WHO, 2002). An important aspect of life in indigenous communities in Africa, Asia, Australia, Canada, Latin America, and any other indigenous location, is the extent to which IK is an attribute of a whole range of human experience at the intersection of the complete body of knowledge, know-how, and practices maintained and developed by a collective of people who use the knowledge in their everyday lives (Charema & Shizha, 2008). Indigenous life and philosophy and the accompanying knowledge are an essential element of the livelihoods of many indigenous communities including traditional healing practices.

In Chapter 15, "Traditional Healing Practices: Conversations with Herbalists in Kenya," *Njoki Wane* examines healing aspect of herbal medicine and converses with herbalists living in seven provinces in Kenya who practice indigenous methods to assess the contribution they have made to contemporary healing practices. *Wane*'s chapter is based on a research project carried out in 2006 where fifty-six herbalists and thirty laypeople were interviewed. The chapter highlights only the conversations with the herbalists who have been practicing for more than ten years. *Njoki Wane*'s research shows that traditional healing practices take into consideration the physical, emotional, mental, and spiritual realities of a person seeking help and that the services provided by herbal doctors are unique and holistic in nature, as they take the patients' total self into consideration during the treatment. *Wane* concludes that IK among the herbalists is based on practices, used in diagnosing, preventing, or eliminating physical, mental, or social disequilibrium, and relies on experience and observation handed down through generations.

In Chapter 16, "To Die is Honey, and to Live is Salt": Indigenous Epistemologies of Wellness in Northern Ghana and the Threat of Institutionalized Containment," *Coleman Agyeyomah, Jonathan Langdon, and Rebecca Butler* present an interpretation of health and wellness from a Ghanaian indigenous perspective which runs contrary to the definitions of the WHO's (World Health Organization) pathological focus. *Agyeyomah, Langdon, and Butler*'s counter argument is a rearticulation of wellness not as a pathological practice, but as a philosophy of life. In this chapter, the authors focus specifically on the dangers that current plans to institutionalize the training of what the WHO calls TMPs in Westernized educational contexts pose not only to the wellness approach most Ghanaians prefer, but also to the underlying philosophy behind this approach. The authors argue that to discuss traditional medicine as merely a practice is to ignore major epistemic issues and only underscores the reason why the systematic approach of orthodox Western medicine is incapable of evaluating the strengths of this alternative path to wellness.

REFERENCES

Abdi, A. (2006). Culture of education, social development and globalization: Historical and current analyses of Africa. In A. Abdi, K. Puplampu, & G. Dei (eds.), *African education and globalization: Critical perspectives* (pp. 13–30). Lanham, MD: Rowman & Littlefield.

Abu-Saad, I., & Champagne, D. (eds.) (2006). *Indigenous education and empowerment: International perspectives*. Lanham, MD: AltaMira Press.

Bargh, M. (ed.) (2007). *Resistance: An indigenous response to neoliberalism*. Aotearoa, New Zealand: HUIA Publishers.

Barnes, R., Gray, A., & Kingsbury, B. (eds.) (1995). *Indigenous peoples of Asia*. Ann Arbor, MI: Association for Asian Studies.

Barnhardt, R. (2005). Creating a place for Indigenous Knowledge in education: The Alaska Native Knowledge Network. In G. Smith, & D. Gruenewald (eds.), *Place-based education in the global age: Local diversity* (pp. 113–133). Hillsdale, NJ: Lawrence Erlbaum.

Battiste, M., & Henderson, J. (2002). *Protecting indigenous knowledge and heritage: A global challenge*. Saskatoon, SK: Purich.

Battiste, M. (2008). The struggle and renaissance of indigenous knowledge in Eurocentric education. In M. Villegas, S. Neugebauer & K. Venegas (eds.), *Indigenous knowledge and education: Sites of struggle, strength and survivance* (pp. 85–92). Cambridge, MA: Harvard Educational Review.

Blaser, M., H. Feit, & G. McRae. (eds.) (2004). *In the way of development: Indigenous peoples, life projects and globalization*. London: Zed Books.

Charema, J. & E. Shizha. (2008). Counselling indigenous Shona people in Zimbabwe: Traditional practices versus western Eurocentric perspectives. *AlterNative: An International Journal of Indigenous Scholarship, 4*(2), 124–141.

Choudry, A. (2009). Challenging colonial amnesia in global justice activism. In D. Kapoor (ed.), *Education, decolonization and development: Perspectives from Asia, Africa and the Americas* (pp. 95–110). Rotterdam, The Netherlands: Sense Publishers.

Choudry, A., & D. Kapoor (eds.) (forthcoming). *Learning from the ground up: Global perspectives on social movements and knowledge production*. New York: Palgrave Macmillan.

Dei, G., B. Hall & D. Rosenberg. (eds.) (2000). *Indigenous knowledges in global contexts: Multiple readings of our world*. Toronto, ON: University of Toronto Press.

Eversole, R. (2005). Overview: Patterns of indigenous disadvantage worldwide. In R. Eversole, J. McNeish, & A. Cimadamore (eds.), *Indigenous peoples and poverty: An international perspective* (pp. 29–37). London: Zed Books.

Grande, S. (2004). *Red pedagogy: Native American social and political thought*. Lanham, MD: Rowman & Littlefield.

Grande, S. & T. Ormiston. (Forthcoming). Globalization as the new colonization: Indigenizing resistance. In D. Caouette & D. Kapoor (eds.), *Beyond development and globalization: Social movement and critical perspectives* (31 pages). Ottawa, ON: Ottawa University Press.

International Fund for Agricultural Development [IFAD]. (2000/2001). Rural poverty report 2000/2001, fact sheet: The Rural Poor. Retrieved from, http://www.ifad.org/media/pack/rpr/2.htm

McMichael, P. (2006). Reframing development: Global peasant movements and the new agrarian question. *Canadian Journal of Development Studies, 27*(4), 471–486.

McNeish, J., & R. Eversole. (2005). Overview: The right to self-determination. In R. Eversole, J. McNeish, & A. Cimadamore (eds.), *Indigenous peoples and poverty: An international perspective* (pp. 97–107). London: Zed Books.

Menon, N., & A. Nigam. (2007). *Power and contestation: India since 1989.* London: Zed Books.

Mignolo, W. (2000). *Local histories/global designs: Coloniality, subaltern knowledges, and border thinking.* Princeton, NJ: Princeton University Press.

Quijano, O. (2000). Coloniality of power and eurocentrism in Latin America. *International Sociology, 15*(2), 215–232.

Rajagopal, B. (2003). *International law from below: Development, social movements and third world resistance.* Cambridge: Cambridge University Press.

Said, E. (1994). *Culture and imperialism.* New York: Vintage Books.

Shizha, E. (2006). Legitimizing indigenous knowledge in Zimbabwe: A theoretical analysis of postcolonial school knowledge and its colonial legacy. *Journal of Contemporary Issues in Education, 1*(1), 20–34.

———. (2008). Globalization and indigenous knowledge: An African postcolonial theoretical analysis. In A. A. Abdi & S. Guo (eds.), *Education and social development: Global issues and analysis* (pp. 37–56). Rotterdam, The Netherlands: Sense Publishers.

Sylvain, R. (2002). Land, water, and truth: San identity and global indigenism. *American Anthropologist, 104*(4), 1074–1085.

Tomaselli, K. G. (ed.) (2007). *Writing in the San/d: Autoethnography among indigenous Southern Africans.* New York: Altamira Press.

Villegas, M., S. Neugebauer, & K. Venegas (eds.). (2008). *Indigenous knowledge and education: Sites of struggle, strength and survivance.* Cambridge, MA: Harvard Educational Review.

wa Thiong'o, N. (1986). *Decolonizing the mind: The politics of language in African literature.* Harare: Zimbabwe Publishing House.

Wane, N. N. (2008). Mapping the field of indigenous knowledges in anti-colonial discourse: A transformative journey in education. *Race Ethnicity and Education, 11*(2), 183–197.

World Health Organization [WHO]. (2002). *WHO traditional medicine strategy 2002–2005.* Geneva: World Health Organization.

PART I

DEVELOPMENT

LEARNING FROM *ADIVASI* (ORIGINAL DWELLER) POLITICAL-ECOLOGICAL EXPOSITIONS OF DEVELOPMENT: CLAIMS ON FORESTS, LAND, AND PLACE IN INDIA

DIP KAPOOR

Who wants to go to the city to join the *Oriyas* (dominant castes/urban outsiders) and do business and open shops and be *shahari* (city/ moderns) if they even give you a chance or to do labor like donkeys to get one meal? Even if they teach us, we do not want to go to the cities, as these are not the ways of the *Adivasi*. We cannot leave our forests (*ame jangale chari paribo nahi*). The forest is our second home [after the huts; meaning added]. You just come out and you have everything you need. . . . My friends and brothers, we are from the forest. That is why we use the small sticks of the *karanja* tree to brush our teeth—not tooth brushes. Our relationship to the forest is like a finger nail is to flesh (*nakho koo mangsho*)—we cannot be separated. That is why we are *Adivasi*.

> *Kondh Adivasi* elder, interview notes, village D,
> Orissa, India, January 2007

INTRODUCTION: ARTICULATING *ADIVASI* EXPOSITIONS OF DEVELOPMENT

Contrary to scholarly attempts at anthropologizing *Adivasi* (original dweller) in the interests of knowledge production about and for purposes of legitimation and formulation of the Indian state's colonial construction (official representations) of the *Adivasi-other*, that is, a construction that assists with the contemporary reproduction of governmentality and the disciplining of *Adivasis* into trans/national market-modernization imperatives (the assimilationist project of "capitalist development"), this chapter attempts to articulate *Adivasi* expositions pertaining to *Adivasi*-state development relations concerning claims to forests, land, and place. State-essentialist representations of *Adivasis* as "tribal others"' (referred to as Scheduled Tribes, or STs, enumerated in the Indian Constitution for purposes of *amelioration*) in need of special protection and rights and as peoples with singular, static, and bounded identities located in segregated spaces (referred to in the Constitution as Scheduled Areas) need to be recognized as a *political* exercise in *exclusive* governmentality and social control (see Kaushik Ghosh in this collection for related discussion). In other words, these representations need to be seen as statecraft designed for

i. the purposes of containment and bureaucratic management of an increasingly vocal *Adivasi* discontent with state-market led development-dispossession (claims on forest, land, and material/cultural space) in India and around the world (Escobar, 1995; Ghosh, 2006; Kamat, 2001; Kearney, 1996; Menon & Nigam, 2007; Rajagopal, 2003);

ii. restricting the *real* import of the politics of these struggles to *Adivasis* alone, thereby reducing the scope of resistance to the development project (assisting with state propaganda or what Jackson [2004] refers to as "justificatory myth-making" [p. 98]) around the widespread acceptance of the need and support for capitalist modernizations and development with the exception of these "backward/special populations") that often, in fact, include a much wider constituency (Kamat, 2001; Kapoor, forthcoming);

iii. enabling the state to weaken the possibility of subaltern coalitions (of *Adivasis* and other related struggles of social groups/classes facing similar socioeconomic circumstances) challenging development dispossession by neoliberal and corporatized-socialist state governments through protective measures for certain groups like STs, often at the expense of other marginal groups like Scheduled Castes, or SCs, who are also located in Scheduled Areas (Baviskar, 1991; Kamat, 2001; Kapoor, 2004); and

iv. representing the *Adivasi* as an *exclusive social segment of the poor* that is in need of humanitarian intervention, state-market-civil society supports in relation to employment/wages, livelihood and a say (participation?) over resource management (e.g., Joint Forest Management schemes) in what the state constructs as a politics *for* inclusion in to the modern mainstream, thereby possibly distorting and attempting to *deradicalize* the politics of some *Adivasi* movements by refusing to recognize an *Adivasi* political subjectivity that problematizes the state-defined *politics for inclusion* by questioning the fundamentals of the assimilationist (forced and "benign" variants) project of modern developmentalism, that is, *Adivasi* political subjectivities exposing a *politics of inclusion*, while asserting *Adivasi* history/agency and spiritual, cultural, epistemic, and political economic ways of being.

While recognizing this politics of state-essentialist representation of the *Adivasi-other*, any attempt to articulate *Adivasi* constructions of *Adivasi*-state-market-development relations here is also predicated on the informed assumption that *Adivasi* can/do *deploy* these very state-essentialisms and associated human rights discourses (*strategic* essentialisms) (Ghosh, 2006; Grande, 2004; Kapoor, 2008) in a complex politics of *Adivasi* struggle, survival, resistance, and regeneration that is seemingly contradictory; for example, *Adivasi* claims can be simultaneously essentialist and nonessentialist appeals that are most times based on the *truth claim/historical reality* of being *original/priori peoples* and/or *durable cultures of difference* or appeals based on "state-sanctioned" essentialist-constructions for Adivasi*ness* in order to avail of related legal-rights recourse available through such state provisions where it might make sense to do so. Relatedly, when it comes to addressing the politics of state-market led development-dispossession, *Kondh Adivasi* in the state of Orissa (the context of this exposition), resort to a multiplicity of strategic maneuvers including outright resistance, partial cooperation, agreement to negotiate limited aspects of a planned intervention, and even cooperation around *Adivasi*-defined "developmentalist claims" (support for activities that have always been undertaken by *Adivasi* but that are now increasingly harder to accomplish, given development-caused ecological and political pressures (e.g., deforestation and legal constructions) of *Adivasi* as encroachers of state land on the state (e.g., state supports for small-scale irrigation or *Adivasi*/community forestry).

This chapter articulates a critical exposition of development emergent from *Kondh Adivasi* political ecological knowledge and learning and knowledge gained through encounters between *Kondh* ways and the *neoliberal development/globalization project* (Escobar, 1995; McMichael, 2006; Menon & Nigam, 2007; Patnaik, 2007) with the view to amplify contemporary *Adivasi* concerns with development as forced modernization or what some authors, writing in reference to various indigenous locations, are referring to as re/colonizations integral to the contemporary projects of development (see Barua in this collection; Kapoor, forthcoming) and neoliberal globalization (see Choudry in this collection; Grande & Ormiston, forthcoming). This is followed by reflection on claims to indigeneity in the Indian context as these relate to the politics of contemporary development dispossession of forest/rural dwellers—deliberations that are pertinent in so far as they might have *material political* implications for *Adivasi* and wider subaltern projects (Ludden, 2005; see Mookerjea in this collection).

These foci are addressed through the author's ongoing research concerning "Learning in *Adivasi* (original dweller) social movements in India" (Kapoor, 2009a; 2009b), a study that employs a participatory critical-interpretive methodology (Kapoor, 2009c) that is partially informed by Tuhiwai-Smith's (1999) insights concerning research methodologies that are decolonizing in design and intent, subsequently emphasizing a local-critical research that is mostly embedded in and constructed from/with indigenous epistemic and political points of origin; that is, *Adivasi's* as coparticipants in the research relationship who "expose development" and in the process educate "outsiders" about the implications of development-dispossession and displacement, neoliberal globalization (the globalization of capitalism), and continued re/colonizations (Choudry in this collection; Grande & Ormiston, forthcoming) of *Adivasi* cultural and material space (Kapoor, 2009d; forthcoming).

In so far as the politics of "outsider-insider" and personal locations/positionality is concerned (Tuhiwai-Smith, 1999), this relationship is not beyond such scrutiny, despite the author/researcher's decade-long relationship with the *Kondh* and *downtrodden caste* (or *Dalits*, who in pejorative terms are referred to as "untouchables") communities in the east coast state of Orissa, India (see Kapoor, 2009c). The pertinence of the implications of what Dirlik (2004) refers to as "epistemic nativism" and the associated "proliferation of racialised world politics" (p.145), that is, romantic notions of a pristine *Adivasi* epistemic/material/cultural space as a living embodiment of a critique of modernity in contradistinction to the ugliness of the modern, rendering *it* impervious to external critiques, while always pertinent at some level, needs to be assessed in relation to the urgencies and compulsions of the *realpolitik* of *Adivasi*-outsider relations and the colonial excesses of the imbrications of the march of modernization, development, globalization, and capitalism (Kapoor, 2009c). Taken to an illogical extreme, such critical propositions on insider-outsider research and wider discursive engagements would silence even minimal prospects (despite the predictable politico-cultural and linguistic distortions of this work, for instance) for any such relationship and subsequently and unwittingly provide a carte blanche (politically speaking) for neo/liberal, conservative and other deliberate modern-assimilationist agendas vis-à-vis indigenous peoples.

KONDH ADIVASI POLITICAL ECOLOGICAL EXPOSITIONS OF DEVELOPMENT: KONDH WAYS, NEOLIBERALISM, AND CLAIMS ON FORESTS/LAND

The People's Manifesto of the *Lok Adhikar Manch* (LAM), currently a coalition of fifteen rural people's movements in South Orissa, India representing mainly *Adivasi* (including *Kondh Adivasis*) and *Dalit* peoples in the state (see Kapoor, forthcoming), states that

> [w]e want to live the way we know how to live among our forests, streams, hills, and mountains and water bodies with our culture, traditions, and whatever that is good in our society intact. We want to be the ones to define change and development for ourselves (in Oriya, the state language—*amo unathi abom parivarthanoro songhya ame nirupono koribako chaho*). We are nature's friends (*prakruthi bandhu*), so our main concern is preserving nature and enhancing it's influence in our lives
> LAM, Field Notes, April 2009

The expressed aspiration can be understood in relation to *Kondh* analysis of state forestry/development (one example) as it pertains to the ways of *Kondhs* and in the words of a *Kondh* leader:

> We are the root peoples (*mulo nivasi*) and the people who dominated us, came here 5000 years ago . . . we fought the British thinking that we will be equal in independent India but today the government (*sarkar*) is doing a great injustice (*anyayo durniti*) . . . the way they have framed laws around land-holding and distribution, we the poor (inclusive reference to several subaltern groups in *Adivasi* locations, including *Dalits*) are being squashed and stampeded in to each other's space and are getting suffocated (*dalachatta hoi santholitho ho chonti*). This creation of inequality (*taro tomyo*) is so widespread and so true. . . . They tell us they want to modernize, make machines and industries for themselves. To do this they are doing forcible encroachment of our land—they are all over our hills and stones. They are coming quietly to our forests and hills and in secrecy they are making plans to dig them up and destroy them (reference to mining development) . . . they are diverting our waterways to the towns for their use. . . . We have become silent spectators (*niravre dekhuchu*) to a repeated snatching away of our resources.
> *Kondh Adivasi* man, Interview notes, village D, January 2007

Another *Kondh* colleague elaborates as follows:

> They introduced fertilizers and pesticides and poisoned our cultivable land and water. They introduced soaps and shampoos, Surf (detergent) and spices, and changed our habits, food habits too. Then they tried to take away our land and forests. They have cut down all the trees from their mountains and forests and now they want to take our forests. They took them away and planted useless trees [referring to Joint Forestry Projects initiated by the developmentalist state]. We had trees from our forefather's times in our forests. They took them away and planted useless trees. Our forefathers never needed to dig holes and plant trees. We never gave trees fertilizer. But they came and spoilt the trees and are now finishing the water . . . now that they don't have the waters of the Rushikulya river at Chatrapur, they want to take the Nanding too. Water from Nanding will irrigate their fertile land and they will enhance their wealth.
> Interview notes, Village T, February, 2008

Commenting on the cultural and customary patterns of *Adivasi*-land/forest relations, another elder underscores the invasiveness of state land classification schemes and revenue demarcations and corresponding notions of individual ownership and the commodification of nature:

> Earlier all these forests and the land area belonged to all the people who lived in the area. In the time of our grandparents we had one common graveyard, we had a common system of sharing (or *bheda* in Saora—another *Adivasi* group living close to *Kondhs*—in relation to sharing of fruits, forest products, meat/game, and land/forest usage) and we had a collective contribution system to support each other (such as *entra* or collective labor on each other's millet/swidden cultivation areas). Land was not assigned to any particular person or family—it was a common claim that goes back to our ancestors. We were together in joys and sorrows.
>
> But since the government's revenue demarcation of land and forests, what belonged to all suddenly got divided in to two *moujas* (areas) of claim and people have started saying, "this is mine and this is mine." They [the *Adivasis* of the neighboring village] are now not allowing us to even cut our trees or bamboo for our use. And we are doing the same. This is not our way.

For the *Kondh*[1] birth is both a biological process of human procreation (visible and visual dimension of life) as much as it is a spiritual emergence of "*Kondh*" from the depths of the earth (invisible and nonvisual spiritual dimension of life). While humans and animals have experienced physical/biological/visual birth *and* spiritual/invisible birth, plants, forests, hills, rivers, streams, and rocks have emerged from beneath the earth and are said to have spiritual/invisible birth. *Kondh* believe that each element in their surroundings is a life and soul entity and that while life is mortal, the soul is immortal. Since the soul is a part of all natural entities, each hill is revered as a *penu* (god) that could simultaneously embody the soul of a *penu* and a dead human being (ancestors). The hilltops are the abode of *penu* manifested in the form of thick vegetation, forests, and forest creatures—these hilltops are revered, protected, and are indicative (by the type/lushness of vegetation/forests) of the qualities of the various *penus*. In choosing a habitation, the health and vigor of the abode of *penu* (e.g., forest cover) suggests a good quality of life/location. *Kondh* shamans have the power of communicating with hills and forests and the respective gods and goddesses of these hills and forests and human ancestors for the purposes of prediction/futurology. Fundamental to this possibility is the understanding that the same type of life runs through human beings and all other natural entities making such intercommunication between all of them possible across time, space, and matter.

This rudimentary exposition of *Kondh* spiritual beliefs underlies and informs forestland use practices and patterns and wider spirituo-ecological interactions. While spirits, for instance, dwell in all parts of a hill (hills and mountains are also considered to be protectors of specific villages), hills are usually divided into, at least, three parts and after *Kondh* settlers have sought permission from the particular *penu* (some of whom are also seen to determine rules restricting hunting, felling of trees, and destruction of vegetation/forests, rules that could be enforced by *penus* through retribution in the manner of disease and tiger attacks, for instance), each part of the hill is usually recognized in terms of

i. a rocky hill top/peak where rocks are literally the specific abode of various *penus* (the most powerful spot), surrounded by a part of the hill usually treed or covered in grass and dense vegetation to keep the area cool so as to create a suitable climate for invocation of the gods by the *Kondhs* when necessary (this area is usually left untouched, except in instances requiring wood for hutment construction/last resort or for festivities—spiritual connotations mostly or ideally ensure forest protection in this zone);

ii. the area below the hilltop (central segment) is used for swidden cultivation (slash-and-burn shifting cultivation of the age-old staple millet/ragi mostly eaten as a porridge/gruel) and is left fallow for two to three cultivation cycles to regenerate before the next use although land pressure (e.g., competition from "outsiders," the state-corporate development process, ecological reserves/conservation measures, etc.) is making this cycle harder to adhere to, encouraging soil erosion and reductions in soil fertility; and

iii. the area on the foothills is literally the area designated for "humans," although several spots are recognized as the abode of various *Kondh* deities as the *Kondhs* share land/forest/trees with the different *penus*. Streams also have their respective deities and while all entities (including the *Kondh* dwellings) are *restricted* to certain spots/areas in relationship with *penus*, vegetation, and water bodies are *unrestricted*.

While swidden cultivation and consumption of millet, peas (cow and pigeon varieties), cereals, and naturally occurring fruit (jackfruit) and vegetable (tubers) have always been and remain the preferred form of *Kond* agriculture/nutrition along with wild game, horticulture and paddy/rice cultivation (on low/wetlands when available) are relatively recent introductions, the former having been encouraged by the tribal development agencies of the state while the latter has entered the area with encroachments by mainly urban Oriya/dominant caste group outsiders. Horticulture allows for consumption and sale in local/nearby markets (*haats*) of banana, mango, jackfruit, guava, and oranges (and citrus fruits in general) and this practice is encouraged by *Kondhs* on transverse plots for commercial value while swiddening is on longitudinal plots (based on rationale pertaining to the relative productivity for each activity). Swidden agriculture and the cultivation and consumption of millet (upper and lower areas) are the safest/tried staple (drought resistant) of the *Kondhs* and usually remains the first/last resort, despite state attempts to ban slash-and-burn agriculture which has been largely discredited by official science as an ecologically unsound practice, contrary to *Kondh* understanding of the same. Big/old trees are protected, as are deified trees/groves like mango, jackfruit, tamarind, and *mahua*. These trees are also respected as the connection to previous *Kondh* generations that preserved them.

Forests surrounding the *Kondh* dwellings are maintained in order to generate timber for housing materials but also because this cover encourages wild game to frequent the area. Trees on the hilltop region are also conserved in order to meet fuelwood and timber needs during times of short supply, while tall trees, even in the slash-and-burn zones, are left standing either for their shade, economic benefit, the tough labor required to fell them, and/or if they are the abode of a *penu* or spirit. Similarly, certain sections of forest are left untouched as they are considered dwellings of the gods/goddesses and felling of trees and holding gatherings in these locations is forbidden. Such sites are selected by the religious heads who *know* through the medium of dreams. Even branches are not broken nor are inappropriate activities such as defecation permitted in these sacred forests. These spiritually informed conservation measures are increasingly threatened by commercial forestry, petty theft, and forest exploitation by non/tribals and outsiders, as some younger *Kondh* members are beginning to doubt the power of *Kondh* ways given their experiences in towns and cities and the knowledge imbibed through a process of formal schooling that is silent about *Kondh* knowledge/ways.

Kondh cremation grounds are located in forested areas that are thick with shrubs, bushes, and trees that are said to be the abode of ancestors and spirits. Felling wood or even gathering wood for fuel from these areas is forbidden, except if it is for a cremation (even this can only be gathered). While abandoned cremation grounds are eventually exploited, no *Kondh* village is ever likely to completely clear and cultivate this land. Even medicinal plants from these forests are seen to be ineffective due to the presence of certain spirits. Predictably then and for these reasons, some of the higher density old growth forests are cremation ground areas.

Kondhs have developed an elaborate base of knowledge and associated practices/taxonomies pertaining to various plant (e.g., medicinal v. edible), tree (e.g., economic value, spiritual value, sociocultural value), and soil classifications (e.g., over six types of soil for instance, with variations assessed based on color, moisture, location, texture, and their respective productivity in relation to cultivation of different crops)—knowledge that informs agroforestry practices, herbal/plant medicinal practices, and conservation and the maintenance of interlocking life cycles of all natural entities (visual/nonvisual). Even in the case of shifting cultivation, site selection is an elaborate process that takes into consideration the ecological state of the hillside (bushy and moist areas with fewer trees are preferred) and whether or not the *penu* (an altar needs to be present) of the specific hill will permit them to proceed after necessary rituals seeking permission have been performed. Once these elements have been assessed, land is normally divided (by the head of the clans and religious heads) proportionally among all the families (bigger families get more land to cultivate given more stomachs to feed and when produce is distributed from collective plots/*rabi* or colder season cultivation, they also receive a proportionate share) of a given clan (and between clans) who wish to cultivate the hillside. Irrespective of clan membership, some land is also normally kept for new members. Lands/hill sites are marked to demonstrate area of cultivation by a given village. Given that shifting cultivation implies allowing for fallow periods (anywhere from two to five years to allow for regeneration and renewed fertility), villages/clans/families often have four or five sites for continued cultivation through the year.

Swiddening and cooperative labor (*entra*) go hand in hand given the labor intensity of slash-and-burn cultivation. While families clear and prepare land, other steps in the process often require the help of others. Participants are usually provided with food and in recent monetized/market times, with cash, which is pooled in a common fund for the village by the head. All participants in a cooperative labor unit usually help each other as and when required, although marketization/individualization of labor for a price/wage through the tribal development authority-initiated horticultural schemes (these require less labor and cannot engage the collective labor unit) is beginning to make a dent in these collective practices and *Kondh* labor rationality/systems (especially among the younger generation).

Kondhs have historically viewed land, water, vegetation, and wildlife as a unified whole that can be used collectively by a clan/village community but over time, individual claims over labor capacity (especially in relation to the cash economy), inherited land (as per the customary system of distributing land to families for swidden cultivation based on need and labor capacity), and products bought and sold through the encroaching market/monetary economy are being exerted and entertained. However, communal claims over the aforementioned natural categories are enforced and individuals can not lay claim to say, for example, fruits that have fallen from trees in the forests. Water bodies, forest vegetation, and wild game are community resources as are any renewable resources such as fruits and the sap from trees (local brew/alcohol), which otherwise might be protected by or is under the watch of an individual for the community. Social restrictions prevent the felling of such fruit-bearing trees in order to ensure the economic strength of the community and possible misappropriation of the same by outsiders.

The *Kondhs* are keenly aware of the politics of *development as dispossession* and *development as violence* as it pertains to claims on forests and land. In the words of an *Adivasi-Dalit* leader of LAM (when addressing the LAM group),

[w]e are the most burnable (expendable) communities and by this I mean we, the *Dalit*, the *Adivasi*, the farmer and the fisherman are always forced to give up what we have, suffer and sacrifice for the sake of what they call

development. Why should the government develop this country at the cost of our way of life? The government and the industrialists and their intellectuals accuse us of being obstacles to the process of development and as enemies of modernization, enemies of progress and enemies of Indian society.

What they mean by this is that we are in the way of their process of exploitation of natural resources for this development. . . . And we know what this means for us—we have people here from Maikanch who know how the state police always act for the industrialists and their friends in government who want to see the bauxite mine go forward in Kashipur against our wishes, even if it meant shooting three of our brothers; we have people here from Kalinganagar where *Dalits* and *Adivasis* are opposing the Tata steel plant and there too, thirteen of us were gunned down by police; we have people here that opposed Tata Steel in Gopalpur and their shrimp culture in Chilika—in all these movements and struggles many people have been killed by the state and industrialist mafias. . . . Why don't you ask the fisherman where all the fish have gone?

It is time we seriously start to think about this destruction in the name of development . . . otherwise, like yesterday's children of nature, who never depended upon anybody for their food security, we will have no option but to go for mass transition from self-sufficient cultivators and forest and fish gatherers to migratory laborers in far away places. After displacement we stand to lose our traditions, our culture and our own historical civilization . . . from known communities we become scattered unknown people thrown in to the darkness to wander about in a strange world of uncertainty and insecurity.

Focus group notes, February 2008

While *Adivasis* constitute some 8 percent (or 80 million plus people) of the Indian population, they account for 40 percent of development-displaced persons (DDPs) and in Orissa (home to sixty-two ST groups numbering some 8 million people or 22 percent of the population), *Adivasis* account for 42 percent of DDPs (Fernandes, 2006). The process of state-restrictions of tribal rights over land and forests dates back to the time of the British imperial government of the 1880s and the various Indian Forest Acts since wherein "British colonialism distorted the land structure, ecology, forest resources and flora and fauna with grave implications for the *Adivasis*" (Behura & Panigrahi, 2006, p. 35) and British rule began the process of detribalization of land and forests, reducing the tribals to encroachers on their own territories.

This trend is being continued in the post-independence period via the Forest Policy of 1952, the Forest Conservation Act (1980), the Wild Life Protection Act (1972), and the Land Acquisition Act, whereby evictions for an undefined larger public interest have been regularized. According to Behura and Panigrahi (2006), some 500,000 people have been displaced by state-corporate development in Orissa between 1951 and 1995. The post-1991 neoliberal turn has exacerbated this trend through policies promoting reservation; leasing of state land to industrialists (the intended/creation of over 300 Special Economic Zones (SEZ) or free-zones is the latest in an unfolding pattern of state-corporate industrial land grabs, with invasive implications for STs in Orissa given that the state boasts 70 percent of India's bauxite reserves located mostly in the Scheduled Areas worth more than twice India's GDP at 2004 prices); the activation of a Wild Life Protection Act that defines the tribal as the enemy of ecology; and demarcations of land/forests for sanctuaries and national parks that exclude tribals (Pimple & Sethi, 2005). When it comes to mining development alone, under the new National Mining Policy (2006) the neoliberal Indian state has already leased one billion tons (of India's estimated 1.6 billion tons) of bauxite to transnational corporations through MOUs (Memorandum of Understanding) (Indian People's Tribunal on Environment and Human Rights [IPTEHR], 2006). The single largest foreign direct investment in Indian commercial history is in coal and iron in the state of Orissa (by Posco Ltd. of South Korea), a US$12 billion project that is being held up by betel leaf farmers and the *Posco Pratirodh Manch*, which includes significant *Adivasi* and *Dalit* participation. Alcan (Canada), Norsk Hydro (Norway), Tatas (India), and Birlas (India) are all casualties (have divested) of the Kashipur bauxite mining venture in the spiritually significant Baphlimali hills and the *Adivasi-Dalit* resistance spearheaded by the *Prakrutik Sampad Surakshya Parishad* (PSSP)—a project that could displace up to 60,000 *Adivasis/Dalits* while diverting 2,800 acres of land (IPTEHR, 2006). *Adivasi* resistance here has been met with a move to denotify (remove from the ST Constitutional Schedules and the rights/protections thereof) 40,000 *Jhodia Adivasis* in Kashipur (makes it easier for the state to acquire land for mining in Schedule Areas) and a combination of bribes and repression (Kapoor, 2009b). As one *Adivasi* leader expressed it:

> They are fighting against those who have everything and nothing to lose. We will persist and as long as they keep breaking their own laws—this only makes it easier for us! That is why, even after the police firing in Maikanch in 2000, over 10,000 of us showed up to oppose the project the very next month.
>
> Focus group notes, February, 2008

The 2006 Scheduled Tribes and Other Traditional Forest Dwellers (Recognition of Forest Rights) Act, or FRA, which was hailed by people's movements, as a victory of sorts in the struggle for *Adivasi* autonomy and sovereignty, as the act appears to recognize the *Adivasi* way of life, is now being viewed in some quarters as yet another "law and 'new welfare model' used by the State to retain it's authority, power and supremacy over resources, alienate people from their land and way of life, and create and sustain capital markets" (Ramdas, 2009, p. 72). While the FRA recognizes community and customary rights to the forest and confers power to the communities to protect forests in accordance with their own modes of conservation, the Ministry of Tribal Welfare and the Forest Department have interpreted these provisions as a license to sanction export/urban market-oriented mono/cash-crop rubber, coffee, and fruit plantations in conjunction with the National Rural Employment Guarantee Scheme (NREGS), whereby tribals are reduced to being a source of cheap labor for these so-called tribal development schemes. According to a government report in relation to the FRA, "Now the tribals can cultivate their lands with dignity and without any fear. Tribals can plant rubber plants, mango, cashew nut, orange, lime or palm oil as per local conditions. The state government would also develop lands in tribal areas and the tribals will be paid daily wages under NREGS program though they are working on their own land" (Ramdas, 2009, p. 69). The "free choice" between palm oil and coffee mono-crop for markets is ironic to say the least, as it goes against the spirit of the FRA (if not and more significantly, *Kondh Adivasi* ways alluded to earlier in this chapter) which emphasizes customary traditions and practices for *Adivasi* purposes which in turn, happen to be geared toward food crops for *Adivasi* families. In fact, the FRA, along with the Panchayat Extension to Scheduled Areas (PESA) is supposed to strengthen the hands of *Adivasi* communities and the local *gram sabhas* to decide on whether or not to implement mono-crop plantations or any other programs that might threaten to displace *Adivasi* production, cultures, ecosystems, knowledge, and ways. Production for food by *Adivasi*, however, is being viewed by Forest Departments (for instance) as "encroachment" and is being met with considerable aggression to evict *Adivasi*/forest-dwelling communities from their homelands across the country (for example, see the World Rainforest Movement bulletin 135, October 2008, for a disturbing analysis of atrocities committed by state agents against *Adivasi/Dalit* women, while allegedly implementing the FRA). In Orissa alone, forest diversion (euphemism for industrial/development land grabs) in the postneoliberal era has doubled between 1991 and 2004, while 26 percent of forest land cleared since 1980 has been after the introduction of the FRA (Wani & Kothari, 2008).

As is the case with the FRA, the climate change agenda is also being utilized to continue the process of *sustainable development dispossession* of *Adivasi* and indigenous people as was made evident in the recent Indigenous People's Declaration on Climate Change (The Anchorage Declaration, 2009):

> [We] Indigenous People challenge States to abandon false solutions to climate change that negatively impact Indigenous Peoples' rights, lands, air, oceans, forests, territories and waters. These include . . . agro-fuels, plantations and market based mechanisms such as carbon trading, the Clean Development Mechanism and forest offsets. The human rights of Indigenous Peoples to protect our forests and forest livelihoods must be recognized, respected and ensured.

The World Bank (2006), however, sees climate change as an investment opportunity that will "assist communities to use forests as a means of moving out of poverty" (p. ix) and suggests that "local ownership offers opportunities to capitalize on forest assets" (p. 13), a notion that is being promoted through the Bank's short-term financing from the BioCarbon Fund to mobilize small/marginal farmers to raise plantations of tree species with high rates of carbon sequestration in their lands, from which they will earn income from carbon credits. According to Ramdas (2009), "The powerful convergence of global climate change policies and neoliberal markets, appears to be an overriding force that is shaping current environment and forestry policy in India. . . . [A]ll initial evidence points towards the displacement of *Adivasi* subjectivities and livelihoods" (p. 72). This is also consistent with some assessments which suggest that promoting market access has become the key neoliberal response to poverty as well; that is, poverty alleviation as instrument/reason for state-market control over and intervention in *Adivasi*/forest dweller's lives, which in turn entails disciplining *Adivasi* who are "presented as inhabiting a series of local spaces across the globe that, marked by the label 'social exclusion', lie outside the normal civil society . . . Their route back . . . is through the willing and active transformation of themselves to conform to the discipline of the market" (Cameron & Palan, 2004, p. 148).

Under these circumstances, a *Kondh* leader's appeal to the government is likely to continue to fall on deaf ears: "We are demanding a place for ourselves—we are questioning the government and asking them to help us develop our land using our ways. . . . If they can help the *shaharis* (urbanites) destroy forests, then they can and should help us to protect it and listen to our story too" (Interview notes, January 2007). The claim on the state (and definitions of citizenship) is to recognize *Adivasi* ways of living and being as a political right and not as an essentialized-inferiority in need of protection and welfare from a self-appointed guardian or paternalistic state. Speaking in relation to forests, for instance, a *Kondh* woman makes this point lucid: "The Forest Department comes and asks us to create a Forest Protection Committee (*jungle surakshya manch*). Protection from whom should I ask? . . . We do not cooperate because they really do not care about the forest! We need to protect the forest from them!" (Interview notes, January 2007, village D). Given that *Adivasi* ways and claims regarding forests/land are still to be given due recognition and space, the political landscape in eastern India is dotted with numerous *Adivasi* (and wider coalitions) struggles/movements pertaining to forests/land, including in Kashipur, Kalinganagar, Lanjigarh, several trans/local struggles in relation to Posco, Singur, Nandigram, and Lalgarh, to name but a few current examples (Kapoor, forthcoming); "political-ecological movements that challenge the development paradigm, and resist statist and capitalist relations that are being violently repressed" (Kamat, 2001, p. 43) by the state.

CLAIMING INDIGENEITY IN INDIA AND THE POLITICS OF DEVELOPMENT DISPOSSESSION

Displacement has become, from the late 1970s, one of the most contested and coercively settled realities of development in contemporary India. *Adivasi*

revolts or movements against displacement have placed the issue of development and dispossession at the heart of political debates in India today. Most
such movements have not been interested in any negotiated rehabilitations
and have only "chosen" to do so after violent repressions orchestrated by the
state and the private corporations involved

Ghosh, 2006, pp. 525–526

Given the level of state repression and market penetration of *Adivasi/
subaltern* space in India (IPTEHR, 2006; Kamat, 2001; Kapoor, 2009d;
Menon & Nigam, 2007), self-definition as *Adivasi* (subaltern claims to being
Adivasi) need to be taken seriously, despite possible strategic deployments of
the same by social groups (e.g., *Dalits*) that are being compelled to scramble
for political, economic, and cultural space in a neoliberal development reality
that re/colonizes rural forests/lands, peoples, and cultures. Discursive analysis
that seeks to unearth these deployments of appeals/claims to being *Adivasi*
(e.g., see Kamat, 2001), while revealing and academically insightful at some
level if not instrumental in advancing a peasant/class political project, could
potentially exaggerate the case for dismissing all claims to being *Adivasi* as
pure caricature/politics or as a strictly *British construction/colonial invention*, and
comes close to disallowing any claims to indigeneity in the Indian context
which, incidentally, is also the position of the hegemonic neoliberal Indian
state. Such propositions also fail to allow for historical (claims of *priority*—e.g.,
being pre-Aryan/Hindu/caste), ethnically (*Kondh* ways/beliefs), and culturally/
place-based claims to being *Adivasi* (as illustrated by several quotes/practices
pertaining to *Kondh* identity/ways and sense of place alluded to in this chapter); nor do they account for the possibility that these struggles are perhaps
as much about a local politics *for Adivasi* peoples as they are simultaneously
about a coalitional counterhegemonic translocal politics of subaltern groups
(Da Costa, 2007; Kapoor, forthcoming) addressing the vicissitudes of contemporary neoliberalism and its re/colonizations. Furthermore, the two
politics need not necessarily become divisive and mutually exclusive as is
suggested by Kamat (2001) and even if this were the case, such self-absorbed
Adivasi movements and *indigenous life projects* by *getting in the way of development* can effectively disrupt (in a given locale) the process of capitalist accumulation and reproduction (Blaser, Feit & McRae, 2004), assuming that this
is the *political basis* of such critiques leveled at localist *Adivasi* struggles (see
Kapoor, forthcoming). The *Kondh* experience in Orissa, however, is indicative of how *Adivasi* and other subaltern groups (e.g., *Dalits*—see Mookerjea
in this collection) recognize the importance of *ekta* (unity) in the face of
state-corporate repression and the divisive deployment of the state-essentialist politics of indigeneity in India. In the words of a *Kondh* woman, "We are
all [referring to *Dalits* and *Saora Adivasis* alike] members of the movement
organization and our struggle is around *khadyo, jamin, jalo, jangalo, o ekta*
(food, land, water, forest, and unity)." Or as a *Kondh Adivasi* leader said,

"The *ucho-barga* (dominant castes and classes) will work to divide and have us fight each other till we are reduced to dust (*talitalanth*) . . . Our *dhwoja* (flag) is *ekta* (unity) and we have to fly it high (*oraiba*). We fly the flag of the people who have lost their land and their forests and who are losing their very roots" (Field notes, village D, January 2007).

In the final analysis, as Kaushik Ghosh (in this collection) observes, while the deployment of the racist logic of *exclusive governmentality* (i.e., the recognition of the negative and separate nature of tribal/*Adivasi* identity) can work to secure state attempts at balkanization and/or containment of development-displaced subalterns, it can also be deployed by *Adivasi* to secure *appropriate* rights/concessions (e.g., land-for-land demands because the state essentializes *Adivasi* as incapable of handling money) through such paternalistic social taxonomies; demands that effectively subvert or interrupt the hegemonic prospects of the original design/intent of *exclusive governmentality* and its ability to assist with the reproduction of the contemporary neoliberal Indian state. When such claims in the name of being *Adivasi* are made with other victims of development-dispossession and global capitalism (e.g., *Dalits* and other forest-dwelling social groups) (Kapoor, forthcoming), that is, with people who live in the Scheduled Areas and similarly have to confront political, economic, and ecological issues as do *Adivasis*, the colonial prospects of neoliberal development, globalization, and compulsory modernization are seriously compromised.

Acknowledgment Note: The author acknowledges the assistance of the Social Sciences and Humanities Research Council of Canada (SSHRC) for this research into "Learning in Adivasi (original dweller) social movements in India" through a Standard Research Grant (2006–2009).

REFERENCES

Baviskar, A. (1991). The researcher as pilgrim. *Lokayan Bulletin, 9*(3), 91–97.

Behura, N., & N. Panigrahi. (2006). *Tribals and the Indian constitution: Functioning of the fifth schedule in Orissa.* New Delhi, India: Rawat Publications.

Blaser, M., H. Feit, & G. McRae. (eds.). (2004). *In the way of development: Indigenous peoples, life projects and globalization.* London: Zed Books.

Cameron, A., & R. Palan. (2004). *The imagined economies of globalization.* London: Sage.

Da Costa, D. (2007). Tensions of neo-liberal development: State discourse and dramatic oppositions in West Bengal. *Contributions to Indian Sociology, 41*(3), 287–320.

Dirlik, A. (2004). Spectres of the third world: Global modernity and the end of the three worlds. *Third World Quarterly, 25*(1), 131–148.

Escobar, A. (1995). *Encountering development: The making and unmaking of the third world.* Princeton, NJ: Princeton University Press.

Fernandes, W. (2006). Development related displacement and tribal women. In G. Rath (ed.), *Tribal development in India: The contemporary debate* (pp. 112–132). New Delhi, India: Sage.

Ghosh, K. (2006). Between global flows and local dams: Indigenousness, locality and the transnational sphere in Jharkhand, India. *Cultural Anthropology, 21*(4), 501–534.

Grande, S. (2004). *Red pedagogy: Native American social and political thought.* Lanham, MD: Rowman and Littlefield.

Grande, S. & T. Ormiston. (forthcoming). Globalization as the new colonization: Indigenizing resistance. In D. Caouette & D. Kapoor (eds.), *Beyond development and globalization: Social movement and critical perspectives* (31 pages). Ottawa, Canada: Ottawa University Press.

Indian People's Tribunal on Environment and Human Rights [IPTEHR] (2006). *An inquiry into mining and human rights violations in Kashipur.* Mumbai, India: IPEHR publication.

Jackson, M. (2004). Colonization as myth-making: A case study in Aotearoa. In S. Greymorning (ed.), *A will to survive: Indigenous essays on the politics of culture, language and identity* (pp. 95–103). New York: McGraw-Hill.

Kamat, S. (2001). Anthropology and global capital: Rediscovering the noble savage. *Cultural Dynamics, 13*(1), 29–51.

Kapoor, D. (2004). Popular education and social movements in India: State responses to constructive resistance for social justice. *Convergence, 37*(2), 210–215.

———. (2008). Popular education and human rights: Prospects for antihegemonic Adivasi (original dweller) movements and counterhegemonic struggle in India. In A. Abdi & L. Shultz (eds.), *Educating for human rights and global citizenship* (pp. 113–128). Albany, NY: SUNY Press.

———. (2009a). Subaltern social movement learning: Adivasis (original dwellers) and the decolonization of space in India. In D. Kapoor (ed.), *Education, decolonization and development: Perspectives from Asia, Africa and the Americas* (pp. 7–38). Rotterdam: Sense Publishers.

———. (2009b). Adivasi (original dwellers) 'in the way of' state-corporate development: Development dispossession and learning in social action for land and forests in India. *McGill Journal of Education, 44*(1), 55–78.

———. (2009c). Participatory academic research (par) and people's participatory action research (PAR): Research, politicization, and subaltern social movements (SSMs) in India. In D. Kapoor & S. Jordan (eds.), *Education, participatory action research, and social change: International perspectives* (pp. 29–44). New York: Palgrave Macmillan.

———. (2009d). Globalization, dispossession and subaltern social movement (SSM) learning in the South. In A. Abdi & D. Kapoor (eds.), *Global perspectives on adult education* (pp. 71–92). New York: Palgrave Macmillan.

———. (forthcoming). Subaltern social movement (SSM) post-mortems of development in contemporary India. In D. Caouette & D. Kapoor (eds.), *Beyond development and globalization: Social movement and critical perspectives* (40 pages). Ottawa, Canada: Ottawa University Press.

Kearney, M. (1996). *Reconceptualizing the peasantry: Anthropology in global perspective.* Boulder, CO: Westview Press.

Ludden, D. (ed.). (2005). *Reading subaltern studies: Critical history, contested meaning and the globalization of South Asia.* New Delhi: Pauls Press.

McMichael, P. (2006). Reframing development: Global peasant movements and the new agrarian question. *Canadian Journal of Development Studies, 27*(4), 471–486.

Menon, N. & A. Nigam. (2007). *Power and contestation: India since 1989.* London: Zed Books.

Patnaik, U. (2007). *The republic of hunger and other essays.* Gurgaon, India: Three Essays Collective.

Pimple, M. & M. Sethi. (2005). Occupation of land in India: Experiences and challenges. In S. Moyo & P. Years (eds.), *Reclaiming land: The resurgence of rural movements in Africa, Asia and Latin America.* London: Zed Books.

Rajagopal, B. (2003). *International law from below: Development, social movements and third world resistance.* Cambridge: Cambridge University Press.

Ramdas, S. (2009). Women, forest spaces and the law: Transgressing the boundaries. *Economic and Political Weekly, 64*(44), 65–73.

Tuhiwai-Smith, L. (1999). *Decolonizing methodologies.* London: Zed Books.

Wani, M., & A. Kothari. (2008, September 13). Globalization vs India's forests. *Economic and Political Weekly*, 19–22.

CHAPTER 3

INDIGENOUS
INCITEMENTS

KAUSHIK GHOSH

THE UNITED NATIONS DECLARATION on the Rights of Indigenous Peoples
(UN, 2007) completed its second anniversary in September 2009. Although
an overwhelming majority of nations are signatories to it, the declaration still
cannot directly or immediately affect policies of most individual nation-states
toward their indigenous populations. This is so primarily because the expected
ambitious pronouncements in the declaration are defanged by the equally predictable fact that the declaration is a nonbinding document.[1] So, in practice,
the declaration cannot support any action that can be found to "dismember or
impair, totally or in part, the territorial integrity or political unity of sovereign
and independent States" (UN, 2007, article 46). The Declaration, therefore,
needs to be read for its more indirect and discursive effects, especially as a
significant formal reference point in contemporary debates on indigeneity. It
certainly brings an obvious visibility to the notion of indigeneity and places it
more firmly within contemporary political vocabulary.

In this essay, I take the opportunity provided by the heightened attention to the concept of "indigeneity" to introduce a set of concerns, which
have not been very firmly pursued by either activist or scholarly commentators on indigenous politics. I focus particularly on two related problems. I start with the problem of indigeneity framed within a distinctly
Euro-American narrative of a progressive liberation, an imagination that
deeply informs the declaration itself. A discussion of this will lead to the
second problem, namely, the tension between indigeneity in the settler
colonies and the postcolony. This second point will be explored through
a discussion of certain features of indigeneity in India in contrast to a
cultural archive of indigeneity that characterizes the settler colonies,
especially those in North America and Australia.

INDIGENEITY: THE NARRATIVE OF PROGRESS

The text of the declaration is a document of political emergence; that is, it is the allegory of the gaining of a rights-based subjectivity by the indigenous. By assuming and producing the movement from a state of "abject being" to that of a "'rights-bearing' subjectivity," the Declaration extends the power of the well-known secular historical narrative (of the nonindigenous) to indigenous populations. But the extensiveness of this power goes well beyond its obvious domain at the United Nations (UN). Critical contemporary scholars of indigeneity, who are otherwise least satisfied with the promises of a rights discourse, seem equally implicated in this reproduction of this narrative of progress. A recent volume edited by de la Cadena and Starn (2007b), *Indigenous Experience Today*, actually suggests a rather more complex and productive framework to approach "indigeneity." The authors explicitly pose the problem of indigeneity as not one of a natural line of progress but rather a complex, overdetermined field. "Reckoning with indigeneity," they observe, "demands recognizing it as a relational field of governance, subjectivities and knowledges that involves us all—indigenous and nonindigenous—in the making and remaking of its structures of power and imagination" (de la Cadena & Starn, 2007b, p. 3). Yet, diverging from this complex and nonteleological approach, their account is haunted by the persistent reappearance of a narrative of progress. At the very beginning of the book, they write:

> A century ago, the idea of indigenous people as an active force in the contemporary world was unthinkable. According to most Western thinkers, native societies belonged to an earlier, inferior stage of human history doomed to extinction by the forward march of progress and history. . . . History has not turned out that way at all. (Today) indigenous peoples have asserted their place in 21st century global culture, economy, and politics.
>
> de la Cadena & Starn, 2007b, pp. 1–2

Similarly, a recent Wenner-Gren Foundation sponsored panel on indigeneity at the American Anthropological Association (de la Cadena & Starn, 2007a), organized by the same authors and participated in by several anthropologists including myself, worked around a panel proposal, which conceptualized the crux of indigeneity as such:

> Although often remaining among the poorest and powerless sectors of their national societies, native groups are *no longer* always among the ranks of the dispossessed and the voiceless. Indigenous filmmakers, writers, artists, and intellectuals have become key interlocutors in debates about memory, history and the representation of native and non-native experience. *Today*, one can be simultaneously modern and Hopi, Maori, or Kayapo, at once indigenous and fully part of the fast-changing, digitalized contemporary world.
>
> de la Cadena & Starn 2007a; emphasis added

Indigeneity emerges here as a trajectory of progress; the trajectory of a movement from an abject past to a liberating present. This relationship of the past
and present represent a definitive "modern" imaginary of time. The past and
present are linked through a telos of progression from death to life.[2] The
temporal duality gets articulated also as a political one. Indigenous people are
not to be thought of as relics of the past but more as participants and producers of the present. This temporal duality is one of dispossession, lack of voice
and powerlessness on the one hand, and agency, voice, and empowerment
on the other.

Why do the indigenous and the modern appear as opposed terms? Why
should indigeneity always evoke an opposite temporality—"fast-changing,"
"modern," "contemporary," and "twenty-first century"? How does this binary
structure invest our discourse of indigeneity even in its most critical moments?

Indigeneity and the concept of indigenous people are particularly pronounced examples of a conflict that deeply informs the operation of the
modern concept of "people." While "the people" signifies the presence of a
political collective that acts with a certain degree of autonomy and agency,
and hence becomes the basis of such powerful instances of political subjectivity as the "nation" or "indigenous people," there is a less explicit but at least
equally formative meaning as well. This is the form in which "the people"
appear as the abject: the embodiment of oppression, subjection, and powerlessness. The potential of sovereignty and autonomy of the people derives
from this other meaning. The two meanings are perfectly interdependent.
Why should people have sovereignty and freedom? It is precisely because
they are (invented as) dominated and disempowered. This is a crucial point
in Agamben's (2000) work:

> Every interpretation of the political meaning of the term "people" must begin
> with the singular fact that in modern European languages, "people" also always
> indicates the poor, the disinherited, and the excluded. One term thus names
> both the constitutive political subject and the class that is, de facto if not de
> jure, excluded from politics. (p. 176)

"Indigenous people" is the name of one the most active inhabitation of
this constitutive opposition. The history of genocides in the settler colonies
marks them as a most dramatic case of the people as abject. And precisely
because of that we see the strong imperative of witnessing their emergence
as subject.

Such an imperative, as it is based on the need to erase the abject and
wretched state, cannot but marginalize it temporally. Indigenous people are
to be grasped not in the time-space of dispossession and displacement but in
the language of political subjectivity and peoplehood. Following this gradi-
ent we find ourselves not only reproducing the opposition "simultaneously
modern and indigenous/Hopi/Maori" but also willy-nilly participating in

the great Euro-American project of seeing liberation in the political subjectivity and temporality of something akin to the citizen. The production of life as part of and in relation to dispossession then feels like something in the past, something that cannot produce political subjectivity and something that at best can only wait to attain the temporality and aesthetics of the citizen through its own dissolution. It is because of this that we find it productive to use the set of oppositions that I have mentioned before: modern and Hopi, indigenous and fully part of the fast-changing, digitalized contemporary world. This, in Agamben's (2000) words is the "fundamental biopolitical fracture" (p. 32) of Western political discourse. We somehow again find ourselves in that modern dichotomy of "bare life" (nature) and "political body" (culture).

I would argue that while superficially it may seem that one is refusing the dichotomy by pointing to their simultaneity, the two elements of the opposition have become too stabilized in our imagination, especially so with the category of indigenous people. While the modern and normative political subjectivity of the citizen gets reproduced as "the fast-changing, digitalized contemporary world" and "twenty-first century life," issues of lives led in close relation to land, dispossession and a noncitizen subjectivity are hardly allowed to go beyond the somewhat predictable and limited politics of unpacking "essentialisms," "exotica," or "colonial inventions."

In the next two sections, I want to take recourse to the colonial and postcolonial histories of indigeneity in India—which comprises the bulk of my primary research—to comment on a different way to rethink indigeneity. By illuminating the political subjectivity of the *adivasi* (indigenous) in Indian modernity, I explore forms of emergence of an indigenous identity, which is particularly difficult to contain within the progress narrative of "abject to subject," which I have discussed so far. I revisit questions of land and political subjectivity in the context of aspects of indigenous or adivasi customary law in India. I find the domain of customary law unavoidable and particularly interesting in discussing the issue at hand.

A QUESTION OF TWO INDIGENEITIES?

Customary law is a discourse of "recognition of exception" that emerged in the interstices of colonial governmentality, and traveled widely through the networks of imperialism and has become an unavoidable aspect of the postcolonial and the transnational (Dirks, 1997; Mamdani, 1996, Scott, 1999). In case of indigenous peoples, the most common form it has assumed is in relation to land tenure, territorial rights, and sovereignty. In India, as in other parts of the world, adivasi peoples' rights to land are largely signified through a sizable grouping of such customary legal exceptions, which aim to protect indigenous rights to land from the superior economic and political power of other dominant groups (Ghosh, 2006; Povinelli, 2002; Singh, 1983; Sundar, 2005; Upadhyay, 2005).

Recognizing them as a form of governance through exceptions, I have elsewhere used the term "exclusive governmentality" (Ghosh, 2006). In the anthropological literature, critical scholars have often pointed to a colonial impetus behind such exclusive governmentalities, especially the form of "governance through native customs" and others have coined the term "the ethnographic state" (Dirks, 2001). Since land protection acts are supposed to be acting on the premise of protection of customs relating to clan kinship and land tenure, they have been typically conceptualized as a colonial primitivist imaginary. This, critics argue, leads to various essentialisms regarding indigenous cultures, unleashing the tyranny of the politics of authenticity and nativism, sometimes leading to the specter of ethnic conflicts, the impossibility of indigenous agency and entrepreneurship, and finally the production of the temporality of a "subject" as opposed to that of a "citizen" (Li, 2010; Mamdani, 1996).

While these are all legitimate fears, I would argue that they are hasty readings leading to the erasure of indigenous agency and ability to force the production of such a discourse of customary indigenous land rights. Two major issues need to be kept in mind here. Firstly, in India, adivasi or indigenous lands had come within the purview of customary law governance from at least the early nineteenth century.[3] This is salient in light of the fact that scholars have claimed that although customary law in the form of Hindu family law was developed in the Indian colony from the early decades of the nineteenth century, in relation to land and territory it developed its form in Africa (Mamdani, 1996). This is factually incorrect and reproduces a dominant representation of the nation in India, where the specificity of the experience of colonialism for adivasi peoples is perennially absent. Customary law in land ownership was the crucial domain of conflict and colonial governance in adivasi history and needs to be treated with great attention. Secondly, and even more importantly, the unfolding story of adivasi customary law allows us to appreciate the complexity and potentiality of this history and discourse of indigeneity. Rather than seeing a story of colonial racism and essentialism leading to the perversion of "homelands" and the tyranny of invented traditions, I would suggest that the domain of customary law in land has proved to be a most significant, and yet unpredictable point of resistance to displacement and dispossession caused by state and neoliberal/global capitalist projects of industrialization. In fact, it is because of the complex play of memory, law, sovereignty, and moralities that is embodied in the customary law regimes in adivasi lands—or exclusive governmentalities—that there has been a persistent and dramatic history of land struggles in India, which today, as mentioned earlier, seems to have become the most important body of resistance to neoliberalism in that country (see Ghosh, 2006 for a particular treatment of this approach).

Adivasi history in colonial India is characterized by the density and frequency of revolts against the state and new classes of landlords and moneylenders brought into adivasi lands by the colonial capitalist state (see, for example, Jha, 1987; Munda & Bosu-Mullick, 2003; Singh, 1983; Sundar, 1998).[4] Although little of that is recognized or taught in the conventional pedagogy of India in American universities, their intensity and importance continues to leave track marks in the historians' text, as can easily be seen through a close examination of the archive created and accessed in the production of the Subaltern Studies. The early texts of the Subaltern Studies group can be interestingly reread as Studies in Indigeneity, a very different project from its usual association with a very differently inflected postcolonialism. But more significantly, these revolts were crucial in *forcing* the production of protective, customary laws in land ownership or the various "tribal tenancy acts," including the Chotanagpur Tenancy Act of 1908 in Jharkhand (Sundar, 2005; Upadhyay, 2005).[5] While these were colonial regimes of recognition, they were yielded by a colonial state taken aback by the fury and persistence of adivasi revolt. Hence these laws are not only the content of regimes of colonial recognition but also the concrete signs of some of the most spectacular achievements of nineteenth-century adivasi struggles against colonial capitalist rule. The mechanism of governmental power in adivasi areas and the particular territorial imaginations of indigeneity embedded in it therefore carry significant marks of this historical agency of indigenous populations in India.

Apart from increasingly comprehensive land protection acts, the exclusive form of colonial governmentality was also established through a model of governance through missionaries and mission stations.[6] Mass-scale adivasi conversions at various periods in various parts of India, while somewhat scandalous for contemporary liberal and nationalist critics of colonialism (and a most explicit target of right-wing Hindutva movement in India), need to be read as signs of active production of their lives by adivasi populations. Missionization led to the formation of a platform, however imperfect and conflict-ridden, of an educated adivasi middle-class in the last decades of the nineteenth century itself. Parallel to the politics of revolt, the politics of this middle class closely matched and sometimes predated dominant Indian nationalism. Compared to the universalized indigeneity based on settler-colony histories, we certainly do not have any moment where the participation of adivasis in capitalist and "contemporary modern life" was not a historical or political possibility. An evolutionary teleology—although often discussed by upper-caste Indian nationalists and by the colonial masters—predicting the inevitable vanishing of the indigenous was not the political reality that determined the lives of nineteenth- and twentieth-century adivasi indigeneity. By the early twentieth century not only were there strong adivasi student unions in places like Ranchi and prominent participation in local elections by members of this middle class, but also the demand for autonomous adivasi provinces like Jharkhand within and without the Indian state had become one of the central forms of adivasi political imagination (Munda & Bosu-Mullick, 2003). In fact, the 1920s and 1930s saw not only the formation of these demands but also the very coinage of the term adivasi—original inhabitants—which energetically came to replace various pejorative precolonial or colonial terms like Kol ("pig eaters"), junglies, tribals, and aboriginals. These significant signs of political agency and consolidation of political gains enabled in the folds of an exclusive governmentality of customary adivasi laws, which itself was secured through the history of rural adivasi revolts, tell a very different story from the narrative of indigeneity that is based on the history of set-tler colonies like North America and Australia. There we hear how a century ago, the idea of indigenous people as an active force in the contemporary world was unthinkable, of policies of relocation and termination in the Eisenhower years, of the lost generation, of official International Labour Organisation (ILO) policies of assimilation as official "global" policy future of indigenous peoples even as late as the 1950s.

Another important aspect of the telos of transnational discourses of indige-neity has been their dependence on the concept of *priority*. Mary Pratt (2007) has quiet insightfully observed, "In English, the cluster of generic descriptors used to refer to indigenous peoples—indigenous, native, aboriginal, first nations—all refer etymologically to prior-ity in time and place. They denote those who were here (or there) first, that is, before someone else who came 'after'" (p. 398). Adivasi indigeneity actually has had less to do with the ques-tion of *priority*, although occasionally that has been attempted. Instead, the governmental recognition of the revolts and struggles against displacement from their land had meant that adivasi indigeneity has been overwhelmingly dependent on an idiom of ethnicity that is pointedly defined by reference to specific histories and relations of exploitation by moneylenders, landlords, the state, and corporations. Apart from the Indian state's resistance to recognizing the adivasis as the only indigenous peoples of India, the relative lack of appeal of the transnational indigenous movement in India may have to do with these differing political idioms of indigeneity at play.

I conclude this short exploration of adivasi territoriality with a discussion of the possibilities hidden in the form of power called exclusive governmen-tality. Specifically, I demonstrate that while adivasi customary laws signify a hard-won space of land and territorial rights which have allowed them to be contemporary and serious contestants in the political stage for more than 150 years now, colonial essentialisms and primitivisms, which are inalienable aspects of the formation of customary law regimes, do not unfold in such a postcolony with the certainty and implications of domination that we may usually associate with them.

DISPLACEMENT, LAND, AND THE POLITICS OF POSTCOLONIAL REHABILITATION

In the case of adivasi exclusive governmentalities, a very important discur-sive beginning lies in the colonial notion of the primitive being incapable of comprehending proper transactions in money. A century before George Simmel's similar early twentieth-century formulation, the colonial state had made this the ideological basis of explaining adivasi revolts against dispossession caused by the violent introduction of private property in land by the colonial state (Banerjee, 2000; Simmel, 1990/1907). Dispossession caused through adivasi indebtedness to moneylenders and landlords—which was the immediate reason for many adivasi revolts—was explained as the consequence of two causes: the adivasi inability to handle the logic of money and the supposedly 3000-year-old internal conflict of India, where an advancing Aryan race (the ancestors of the Hindus) conquered the Aboriginal races (the ancestors of the adivasis). Such racial and historical explanations then implied that the natural identity of adivasis derived from their belonging to the land. This land-based nature of adivasi authenticity then could become the basis of a politics of paternalist protection, which was then enshrined into the various customary laws regarding adivasi land rights.

While this imagination certainly is an example of colonial essentialism informing the body of colonial customary law, I have already argued that it is the intensity and persistence of adivasi revolts threatening the continuation of the colonial state that had brought the very question of the need for customary land rights to the fore (also see Ghosh, 2006). More ironically, it is the afterlife of such laws that provides us with a glimpse of how the domain of customary law and exclusive governmentalities can produce forms of political agency that deeply destabilize the surety with which the state and capital attempts to appropriate adivasi lands.

The idea that adivasis are incapable of living in the temporality of the market and money exchange and therefore can only survive on the land comes into play repeatedly in various postcolonial contexts of adivasi displacement due to construction of dams, factories, or industrial parks. In adivasi and non-adivasi consciousness, the discourse of adivasi incapability of dealing with money leads to a series of oppositions to the idea of providing monetary compensation to the displaced. This has become a kind of postcolonial common sense in India, when it comes to questions of rehabilitation, with the implication being that such rehabilitation can only take the form of land-for-land and not money-for-land. I have written in greater detail about the specific role of this common sense in the landmark movement against the Koel-Karo hydroelectric project in India (Ghosh, 2006). What started as a racist logic in a system of exclusive governmentality prescribed for tribal regions of colonial India has come to produce a mode of resistance that the modern state had hardly bargained for. In fact, not only in colonial paternalist protection or in adivasi deployments has this essentialized, negative identity centered on the adivasi inability to grasp the universality of money and modernity have had an important role to play, but also increasingly in nation-state governance itself. In 1998, the Supreme Court of India passed a judgment whereby all development projects that threaten to displace adivasi populations must work with a land-for-land rehabilitation plan as opposed to rehabilitation through monetary compensation. Today, this has become standard for all projects displacing adivasis; as far as possible, rehabilitation cannot be done in the form of monetary compensation but only through land-for-land exchanges. Currently, in the aftermath of some of the most robust non-adivasi peasant struggles in India against attempts at land acquisition for the building of massive Special Economic Zones (SEZs), the state is finally drafting a new rehabilitation policy where land-for-land is the general requirement in cases of displacement of all rural populations.[7]

What then began as an act of exclusive governmentality—the recognition of the negative and separate nature of tribal/adivasi identity—became subversive to the project of governance in general. To incorporate the subjects of alterity, to make them continuous with the time of the modern, the state now has to include this alterity as a conceptual basis of its actions. In the process, the very context of the nation-state—its legality and neoliberal land-acquisition projects—is fundamentally redrawn. Even more importantly, this reveals that the functioning of this "exclusive" principle of governmentality may at times interrupt the hegemony of the state and its projects of governance. If the original object of an exclusive governmentality, in the case of adivasis, was the prevention of revolts and other acts that challenge the legitimacy and the functioning of the state, in its reproduction by adivasi antidisplacement movements, we have the inadvertent result of the subversion of the project of development of the contemporary state. Not all situations would produce such subversion, but its very possibility, not to mention its actual materialization in different sites, becomes intrinsic to the process of governmentality and the career of the state.

To speak of indigeneity in the Indian context, then, is to speak of a complex interwoven history of adivasi movements and exclusive governmentality, including customary laws. Together they compose an unstable discursive terrain within the history of national modernity in India. Adivasi persons as well as adivasi struggles overwhelmingly inhabit this space of a fairly old and distinctly dynamic indigenous political world that holds much promise with regard to the possibilities of a politics of indigeneity. Such worlds are not receding residues of an original, authentic "indigeneity." Rather, they are the products of a long struggle in the interstices of a governmentality involving adivasi imaginaries. The latter are deeply marked by this struggle but as they inhabit and deploy this power they also rework it to produce unanticipated dilemmas for the projects of reproduction of state and capital.

The above modality works in twenty-first-century neoliberal capitalism as well. The entire question of land acquisition through the employment of the Land Acquisition Act of 1894 is under some question now. While SEZs multiply, so has the discontent against them. Indigenous adivasi politics of land and territory—although clearly inseparable from colonial governmentality of customary laws in adivasi areas—has managed to produce and keep alive the possibility of mass protests and denaturalization of the discourse of development through displacement.[8]

There are two definite advantages gained through this navigation of the space of adivasi India. Firstly, it forces a serious questioning of the "problem of temporality" of indigeneity as belonging to the modern telos of the political as a movement of a "people" from an abject past to a liberating present. In the process, it also opens up the associated dichotomies of the passive victim and the active subject. An examination of adivasi indigeneity fundamentally destabilizes the operation of the modernist series past/present, passive/active, unfree/free. Specifically, the signs of political agency and consolidation of political gains enabled in the folds of an "exclusive governmentality" of customary adivasi laws, which itself was secured through the history of rural adivasi revolts, tell a very different story of indigeneity than that based on the history of settler colonies like North America and Australia. In the latter case, we hear how a century ago, the idea of indigenous people as an active force in the contemporary world was unthinkable, of policies of reloca- tion and termination in the Eisenhower years, of the lost generation, and of official ILO policies of assimilation as official postwar policy future of indigenous peoples. Constituting a very different trajectory of indigeneity, adivasis have been a potent political force throughout Indian modernity. They won customary law protection and some form of territorial sovereignty quite early in the nineteenth century and had developed a middle-class and student unions by the late nineteenth and early twentieth century. Today the most important resistance to neoliberal capitalism has emerged around land struggles, including a large number of adivasi ones, which have been facilitated by the tradition of movements against land-dispossession in the last three decades, a tradition which is overwhelmingly adivasi in location, content and characterization.

Secondly, and related to the first, the invoking of adivasi indigeneity in relation to a transnational indigeneity allows for a richer exploration of transnational indigenous politics beyond the obvious telos of "the abject to the subject." This is especially relevant for highlighting the problem of under- standing the relation of indigeneities of the settler colony and the postcolony. The discourse of transnational indigeneity contains an underexplored tension between those of settler-colonial states and postcolonial ones, which Richard Lee has termed Indigeneity 1 and Indigeneity 2 (Lee, 2006).[9] I have suggested in this chapter that on various key formulations the settler-colonial and the postcolonial diverge significantly on the question of indigeneity. This is par- ticularly true in relation to the problem of temporality that I have discussed in this chapter. When not fully recognized, as in large parts of the indigeneity scholarship and the transnational indigenous movement facilitated through the United Nations, both of which are dominated by the imagination of present-day movements of the indigenous in the polities derived from settler colonies, there is an immense possibility of erasing the historical robustness and other achievements of the indigenous throughout colonialism and in the postcolonial world.

For indigeneity to realize its potentiality, as Mary Louise Pratt (2007) has recently observed, we need to ask, "How might the fertility or potency of thinking and knowing through (i.e., by means of) the indigenous be appreciated?" (p. 400). This is the crux of the matter, I think. For indigeneity to be a transformational engagement, which destabilizes even the very meaning of "politics," it needs to be read for the new vocabulary of life and the politics that it enables. It should not be contained within the contours of a definitive legal status by being made to stand in as the surrogate for an unexamined progressive politics. In this chapter, I have tried to map an alternate archive of indigeneity that poses critical questions of the telos of our discourses of politics, rights, and liberation. Indigeneity's radical potential, after all, lies in such alternate archives or configurations, the very assembling of which necessarily demands a realignment of the coordinates of the episteme and ontology of the nonindigenous itself. It is with the hope that, in this essay, I have initiated the beginning of a tentative answer to Mary Pratt's question above that I shall end here.

REFERENCES

Agamben, G. (2000). *Means without ends: Notes on politics.* [Trans. Vincenzo Binetti and Cesare Casarino]. Minneapolis: University of Minnesota Press.

Banerjee, P. (2000). Debt, time and extravagance: Money and the making of "primitives" in colonial Bengal. *Indian Economic and Social History Review*, *37*(4), 423–445.

de la Cadena, M., & O. Starn. (eds.) (2007a). Proposal for the Wenner-Gren sponsored panel, "Indigenous Experience Today." 106th American Anthropological Association Annual Meeting, November 29.

——— (2007b). *Indigenous experience today.* Oxford: Berg.

De Sa, F. (1975). *Crisis in Chotanagpur.* Bangalore, India: Redemptorist Publications.

Dirks, N. (1997). The policing of tradition: Colonialism and anthropology in South India. *Comparative Studies in Society and History*, *39*(1), 483–503.

——— (2001). *Castes of mind: Colonialism and the making of modern India.* Princeton, NJ: Princeton University Press.

Ghosh, K. (2006). Between global flows and local dams: Indigenousness, locality, and the transnational sphere in Jharkhand, India. *Cultural Anthropology*, *21*(4), 501–534.

Jha, J. C. (1987). *The tribal revolt of Chotanagpur (1831–1832).* Patna, India: K. P. Jayaswal Research Institute.

Karlsson, B. G. (2003). Anthropology and the "indigenous slot": claims to and debates about indigenous peoples' status in India. *Critique of Anthropology, 23*(4), 403–423.

Lee, R. B. (2006). Twenty-first century indigenism. *Anthropological Theory, 6*(4), 455–479.

Li, T. M. (2010, February). Indigeneity, capitalism, and the management of dispossession. *Current Anthropology, 51*(1), page numbers not known.

Mahato, S. (1971). *Hundred years of Christian missions in Chotanagpur since 1845.* Ranchi, India: Chotanagpur Christian Publishing House.

Mamdani, M. (1996). *Citizen and subject: Contemporary Africa and the legacy of late colonialism.* Princeton, NJ: Princeton University Press.

Munda, R. D., & S. Bosu-Mullick. (eds.) (2003). *The Jharkhand movement: Indigenous people's struggle for autonomy in India.* International Work Group for Indigenous Affairs, Document 168.

Povinelli, E. (2002). *The cunning of recognition: Indigenous alterities and the making of Australian multiculturalism.* Durham, NC: Duke University Press.

Pratt, M. L. (2007). Afterword: Indigeneity today. In *Indigenous experience today,* ed. M. de la Cadena and O. Starn, pp. 397–404. Oxford: Berg.

Roy, S. C. (1995/1912). *The Mundas and their country.* Ranchi, India: Catholic Press.

Scott, D. (1999). *Refashioning futures.* Princeton, NJ: Princeton University Press.

Simmel, G. (1990/1907). *The philosophy of money.* London: Routledge.

Singh, K. S. (1983). *The Birsa Munda revolt.* Delhi: Oxford University Press.

Sundar, N. (1998). *Subalterns and sovereigns: An anthropological history of Bastar, 1854–1996.* New Delhi: Oxford University Press.

———— (2005, October 8). Laws, policies and practices in Jharkhand. *Economic and Political Weekly,* 4459–4462.

United Nations [UN] (2007). Declaration on the rights of indigenous peoples. Retrieved from, http://untreaty.un.org/cod/avl/ha/ga_61-295/ga_61-295.html.

Upadhyay, C. (2005, October 8). Community rights in land in Jharkhand. *Economic and Political Weekly,* 4435–4438.

AGAINST THE FLOW: MAORI KNOWLEDGE AND SELF-DETERMINATION STRUGGLES CONFRONT NEOLIBERAL GLOBALIZATION IN AOTEAROA/ NEW ZEALAND

AZIZ CHOUDRY

> [I]f we are to combat the transnationals, if we are to combat globalization as we must, then we will only do so successfully if we keep it in the context of that centuries-long culture of colonization. . . . We will also only succeed if we attempt as non-colonizers to reclaim one of the most precious things the colonizers took from us, that is the ability to think, to dream in our own words, to find a way to demolish the house that the colonizers built with the tools and words that are ours. (Moana Jackson, 1999, p. 105)

INTRODUCTION

Much has been written about Indigenous Peoples' responses to neoliberal globalization.[1] But the intellectual contributions of Indigenous thinkers and activists towards conceptualizing and contextualizing this process in a much longer history of colonialism and resistance are often overlooked. Continued assertions of self-determination and demands for decolonization by many

Indigenous Peoples are a rich source of theory and critique of both capitalist economic systems and the colonial nature of the state itself. As I argue elsewhere (Choudry, 2009), transnational mobilizations of Indigenous Peoples are one facet of a multilayered, multifaceted, complex struggle, linking to local, concrete struggles against both state and capital, and regional initiatives. Many contend that the transnational corporations (TNCs) are new colonial forces, newer versions of the colonial chartered corporations, such as the British East India, Hudson's Bay, or New Zealand Companies (Jackson, 1999, 2007a; Kelsey, 1999). This chapter highlights Indigenous (Maori) analyses and action against neoliberal capitalism, contextualizing this phenomenon historically, emphasizing how neoliberal theory and practice commodify all things; exalt individualism over collective obligations, reciprocity, and community; are fundamentally predicated on exploitation of people and nature rather than an interrelationship which ensures their mutual survival, and embody a colonial mindset. Burgmann and Ure (2004) argue that

> the practical critique of neoliberalism embodied in indigenous people's resistance to their incorporation into the global market is one informed by an often acute recognition of not only the global dimensions of such resistance but also an acknowledgement of anti-imperialist struggles stretching back many hundreds of years . . . the contributions of indigenous struggles for self-determination . . . have enabled non-indigenous groups and movements to root their critique in an anti-capitalist perspective that emanates from non-Western sources. (p. 57)

Beginning with an overview of colonialism and neoliberalism in Aotearoa[2]/ New Zealand, I discuss the context for Maori resistance to globalization. I pay particular attention to Maori critical knowledge production regarding connections between colonization and neoliberal ideology and practice by focusing on (a) the struggles over traditional knowledge in an era of intellectual property rights paradigms; (b) the tensions with non-Maori environmentalism and nongovernment organizations (NGOs); and (c) the examples of cultural forms of resistance. I draw mainly upon analyses from the 1990s at a time when ongoing Maori struggles for self-determination and growing resistance against several key institutions and processes advancing neoliberal globalization gained some momentum. These mobilizations prefigured the rise of what is sometimes called the "antiglobalization movement" or the "global justice movement." Targets included the General Agreement on Tariffs and Trade/World Trade Organization (GATT/WTO), the Asia-Pacific Economic Cooperation (APEC) forum, the failed Multilateral Agreement on Investment (MAI) negotiated among the Organization of Economic Cooperation and Development (OECD) governments, and the Asian Development Bank (ADB). Much of the analysis I draw upon was first published in small activist/NGO publications and has thus far remained largely outside the realms of scholarly literature. As an activist/organizer, researcher, and editor of several such publications, I also draw on close collaboration with Maori sovereignty activists in confronting capitalist globalization.

Since colonization, Maori have consistently challenged the supremacy and legitimacy of the colonial state. Building on Maori critiques of neoliberalism, this chapter suggests that the past twenty-five years of free market policies must be understood in the context of an ongoing colonial occupation and commodification of Maori lands and resources on which the New Zealand nation-state is based. Maori put forward conceptual tools for understanding and resisting global capitalism, and implicitly or overtly critique some dominant strands of "civil society" challenges to neoliberalism. For example, many "white progressive economic nationalists" (Choudry, 2009, p. 99) in global justice movements still advocate alternatives calling for retooling the nation-state, democratic reforms, and greater state intervention in the economy as they frame the "problem" in a way which ignores the fundamentally colonial nature of the nation-state, and Indigenous Peoples' rights. Furthermore, the politics of knowledge production and academic citation raises questions about whose voices are heard, validated, and overlooked. Despite cogent critiques of free market capitalism anchored in older struggles against colonialism and assertions of self-determination, many dominant NGO narratives and campaign platforms about globalization still overlook or sideline these analyses. Indigenous Peoples in the global North face specific challenges in both local and transnational mobilizations. The "justificatory mythmaking" (Jackson, 2004, p. 98) apparatus of liberal democratic governments in settler-colonial states like Aotearoa/New Zealand is of a different, and perhaps more difficult order to expose and confront than Third World governments, which are more readily seen as unjust and undemocratic in international arenas.

COLONIZATION, NEOLIBERAL GLOBALIZATION, AND AOTEAROA/NEW ZEALAND

In November 1991, Syd Jackson, the late Ngati Kahungunu/Ngati Porou trade unionist and Maori sovereignty activist, addressed a packed public meeting on GATT and free trade in Christchurch.[3] Arguing that free market economics and free trade were nothing new for Maori, he reminded us that the sense of loss of sovereignty that many non-Maori New Zealanders felt, as privatization and deregulation delivered the economy into TNC control, was something that Maori had experienced far more deeply, for generations. Jackson clearly and articulately wove together the threads of colonialism, neoliberalism, Aotearoa/New Zealand's history, and Indigenous Peoples' resistance in a way both firmly located in a local context, but cognizant of global capitalist relations.

The carefully crafted image of a clean, green, antinuclear, socially progressive Western democracy, enjoying model race relations, masks the ongoing colonization of Aotearoa. Since the signing of the Treaty of Waitangi in 1840 between Maori and representatives of the British Crown, Maori sovereignty has been denied. Introduced diseases, wars, dispossession, and the profound disruption of Maori societies had threatened their very survival. In the late nineteenth century, Maori were pronounced a dying race, but now comprise around 15 percent of the total population. The treaty affirmed Maori *tino rangatiratanga* (sovereign right of self-determination) and allowed *Pakeha* (European) settlers to govern themselves. Yet successive New Zealand governments have expediently interpreted the treaty as a cession of Maori sovereignty.

Since the 1980s, Aotearoa/New Zealand has undergone the most radical free market reforms of any OECD country. With a small population (slightly over four million by 2008), this was indeed, as Jane Kelsey (1995), a prominent critic of this process called it, the "New Zealand Experiment—a world model for structural adjustment" (p. 1). Alongside the domestically driven free market experiment, successive New Zealand governments put themselves at the extreme edge of economic globalization ideology. For many, the New Zealand Experiment arrived like a tsunami. The extreme nature of the market reforms unleashed by the (center-left) Fourth Labour Government on an unsuspecting population in 1984, and the relatively high proportion of Maori to the total population (by comparison to that of Indigenous Peoples in Australia, Canada, and the United States) adds another dimension to understanding the dialectics of colonization and resistance in Aotearoa/New Zealand.

Rapid trade and investment liberalization accompanied extensive corporatization and privatization and radical public sector restructuring. Tax cuts for the rich accompanied welfare cuts for the poor. The labor market was deregulated and trade unions decimated (Coney, 1997; Hyman, 1994; Jesson, 1999; Kelsey, 1995). Between 1988 and 1993, Aotearoa/New Zealand led the world in sales of state-owned assets to overseas investors, often at bargain basement prices. Most of its productive, financial, energy, retail, transport, media, and communications sectors are now in the hands of TNCs that have sucked huge profits out of the country (Jesson, 1999; Kelsey, 1995, 1999; Rosenberg, 1993). UNCTAD's (United Nations Conference on Trade and Development) World Investment Report 2000 described New Zealand as the most transnationalized OECD economy (UNCTAD, 2000). The National (conservative) Party, in government from 1990 to 1999, vigorously continued the reforms. The country went into recession with the highest unemployment since the 1930s depression. Nearly 76,000 jobs were lost in manufacturing between 1986 and 1992 (25 percent of all manufacturing jobs)—many directly related to aggressive tariff reduction. Most new jobs that were created were part-time, casual and poorly paid (Coney, 1997; Hyman, 1994; Jesson, 1999; Kelsey, 1995). Kelsey (2002) observed, "The cumulative economic cost to New Zealand of embracing the global free market has paralleled that of many poorer countries" (p. 39).

The social damage was appalling. A 1995 Joseph Rowntree Foundation (JRF) report revealed that over the previous fifteen years, the gap between rich and poor had increased much faster than in any comparable industrialized country. Underfunding, user-pays health and education charges, and the introduction of market rentals for state house tenants severely impacted the poor. Feminist writer Sandra Coney (1997), commenting on health reforms, argued that "the operation was successful but the patient died" (p. 157). The number of New Zealanders estimated to live below the poverty line rose by at least 35 percent between 1989 and 1992. In 1993, charity-run food banks provided NZ $25 million worth of aid. By 1996 about one in five New Zealanders—and one third of all children—were considered to be living in poverty. Youth suicide rates were alarmingly high (Kelsey, 1995, 1999).

While many were taken off-guard by the speed and comprehensiveness of the reforms, some of the strongest challenges to the neoliberal tsunami came from Maori, through legal challenges, direct action, and other methods. Prior to corporatization and privatization, land and resources appropriated from Maori became "public" or "state-owned" assets. Many were outraged that now that these assets were being bought and sold in a global marketplace, they were even further out of reach.

Maori educationalist Graham Smith (Ngati Apa, Te Aitanga A Hauiti) (1993) explains that "[h]istorically the same processes of commodification were used by Pakeha to access Maori land. This was achieved through the individualisation of Maori land titles i.e. to commodify or 'package up' what were collective or group held titles into individual holdings in order to facilitate their sale to Pakeha under Pakeha rules and custom" (p. 7). According to academic/activist Maria Bargh (Te Arawa, Ngati Kearoa, Ngati Tuara, Ngati Awa) (2007a), neoliberalism "demonstrates . . . a translation of many older colonial beliefs, once expressed explicitly, now expressed implicitly, into language and practices which are far more covert about their civilizing mission" (p. 13). Leonie Pihama (Te Atiawa, Ngati Mahanga, Nga Mahanga a Tairi) (1999) argued:

> We have been privatized. We have been put into private homes, on our own private lives, with no accountabilities, with limited support. With our private mortgage and private bank accounts and our private individual way of life. We are in our being and thinking privatized. Little wonder that there is little resistance to the privatization of everything else in this country. (p. 5)

Health, education, and employment statistics, and arrest and imprisonment rates speak of the impact of colonization, racism, and the imposition of alien values and laws on Maori, who were also disproportionately represented in production, transport, equipment, and laboring jobs which were badly hit during the economic reforms (Kelsey, 1995; Rosenberg, 1993). Cuts to welfare hit Maori hard. Urban and rural poverty spiraled upward.

Market ideology has underpinned "Maori affairs" policy of successive governments. Poverty, cultural loss, social and economic conditions—and the fact that the government's approach to settling the treaty claims is presented as the only door open to Maori—have placed enormous pressure on many *iwi* (tribes/First nations). In 1994–1995 the government tried to settle all outstanding treaty claims for a total of NZ $1 billion (the "fiscal envelope"). These cash-for-sovereignty deals were rejected by Maori across the country. Yet successive governments have pressured many iwi into "full and final" settlements (Bargh, 2007a; Kelsey, 1999). A 1992 deal purported to settle Maori fisheries claims pushed aside traditional "spiritual, collective, reciprocal, perpetual and sustainable" (Kelsey, 1995, p. 320) relationships to fisheries by imposing a privatized quota-rights model.

During the campaign against the MAI, Maori challenged the government's mandate to negotiate another international agreement, which would give foreign investors enforceable rights over the resources they were fighting to control. An April 1998 anti-MAI *hikoi* (march) gained support from Maori and non-Maori. Seven hastily organized official Maori consultations, which many condemned as a cosmetic afterthought, rejected the agreement. Maori Members of Parliament from different parties criticized the substance of government commitments to the MAI and the way in which the Ministry of Foreign Affairs and Trade had treated Maori in relation to the agreement. Maori resistance to neoliberal globalization continued through 1999 when the New Zealand government chaired the APEC meetings, alongside state/private sector-supported Maori/Indigenous forums. Since then, opposition has targeted bilateral trade and investment agreements such as the New Zealand-Singapore Closer Economic Partnership deal signed in 2000 (Aotearoa Educators, 2000; Bargh, 2007b), as well as what are arguably local manifestations of commodification and corporate colonialism, such as genetic engineering (GE) and cultural appropriation of Maori cultural practices and traditional knowledge (see Maniapoto & Mills, 2005; C. Smith, 2007). In turn, Maori activists have often borne the brunt of highly politicized police and security intelligence agency surveillance and repression (Choudry, 2005; Kelsey, 1999).

While some Maori chose corporate paths, some embracing free market capitalism as a way to try to bypass the state, others firmly resisted neoliberal globalization. Their resistance and insistent assertions of self-determination serve as a warning that a mere reversion to a "kinder" version of social democracy—still based on invasion, injustice, dispossession, and denial—is not a sustainable or just alternative to the free market agenda. Maori had long mobilized to fight for the survival of their language, and since the 1960s had pushed to set up educational initiatives based on Maori philosophy and values—*kohanga reo* (language nests for preschoolers), *kura kaupapa maori* (total-immersion schools), and *whare wananga* (tertiary institutions) (L. T. Smith, 1999, 2006; Walker, 1990). These have been important sites of politicization, learning, and knowledge production in relation to broader struggles (L. T. Smith, 1999; Walker, 1990). Maori struggles brought forth rich, sophisticated analyses, which located capitalist globalization in an anticolonial framework, drawing upon a worldview that has been in continual confrontation with the processes of commodification, privatization and expropriation, individualism, and imposition of property rights, which started with the New Zealand Company in the mid-nineteenth century (Jackson, 1999, 2007a; G. H. Smith, 1993; L. T. Smith, 1999). This worldview is sourced in a cosmology inherent in which are several interwoven fundamental principles and processes. These are (a) *mana atua,* the interdependent processes, cosmic principles and values distilled from the evolutionary creation of life; (b) *mana tangata,* the authority of human political organization derived from adherence to those principles; and (c) *mana whenua,* the relationship between people and nature that ensures their mutual survival (Kelsey, 1999; M. Smith, 1998).

While refusing to honor the treaty, governments have made far-reaching binding international commitments that lock in domestic neoliberal reforms, without consulting Maori or non-Maori (Jackson, 1995; Kelsey, 1995, 1999; National Maori Congress, 1994). Maori challenged the government's right to commit to international trade agreements on their behalf. In November 1994, the pan-iwi Maori Congress rejected the Crown's ratification of GATT, exempting member tribes from its provisions. It criticized the government for overstepping its treaty responsibilities and democratic mandate by not seeking the public's consent before signing.

> The Crown has not only neglected its Treaty of Waitangi responsibilities to the Iwi Maori treaty partners, but it has violated the very principles of democracy by not adequately informing the public of the pros and cons of the GATT agreement and by not seeking the consent of New Zealanders before signing and ratifying an international agreement which has such widespread and direct consequences on the lives and livelihoods of individuals and communities. (National Maori Congress, 1994)

Jackson (1994) argued that "[i]f GATT is seen by many Pakeha as a delimitation of Crown sovereignty, it is similarly seen by many Maori as a denial of rangatiratanga" (p. 5). Ngati Pikiao lawyer/activist Annette Sykes challenged potential overseas investors and development bankers at a press conference held during the May 1995 ADB's annual meeting in Auckland: "It's about time you sat down and talked to us because the present illegal government has no warrant to deal with resources, neither for the past, nor the present, and certainly not for the future" (transcript of press conference, May 4, 1995). The year 1995 saw several Maori land occupations/reclamations asserting sovereignty in the context of the government's unquestioning embrace of the global free market economy.

Graham Smith (1998), Jackson (1994), Bargh (2007b), and others view the Treaty of Waitangi as a structural impediment to neoliberal globalization, noting interconnections between the logic of treaty settlement policies and the government's international trade and investment commitments. Apologists for globalization often talk about the need to remain attractive to foreign investors, to provide certainty and avoid confusion arising from the treaty for actual or potential foreign investors. Under the provisions of free trade and investment agreements, which New Zealand signs, any preferences afforded Maori under the treaty may attract claims that they are barriers to free trade and investment.

NEOLIBERALISM, THE NEW ZEALAND EXPERIMENT, AND MAORI CRITICAL KNOWLEDGE AND LEARNING THROUGH RESISTANCE

Pihama (1999) connects the contemporary ideology and practice of privatization, the undermining of collective sovereignty exercised by Maori, and resistance to older anticolonial struggles:

> The privatisation agenda in this country did not start with the 1984 Labour government or the MAI or the GATT. . . . Privatisation began with the various Native land acts that saw the individualism of land title from *whanau* [extended family], *hapu* [subtribe] and *iwi* [tribe/First Nation] *kaitiakitanga* [guardianship], so that land could be individually owned, privately owned and therefore available for private and individual gain. Privatisation began with the confiscations of thousands of acres of land by the State that saw Maori dispossessed and the State with an ability to divide lands in order to package and sell them . . . It is that exact privatization and individualism that saw Taranaki remove surveyors' pegs [in the nineteenth century], plough the land, because our *tupuna* [ancestors] knew that the segmenting of our lands was a fundamental part of the fragmentation of us as a people. (p. 4)

Fifteen years into the New Zealand Experiment came Ngati Awa/ Ngati Porou educationalist Linda Smith's (1999) widely acclaimed book, *Decolonising Methodologies: Research and Indigenous Peoples*. As Linda Smith (2006) noted later, "In the neo-liberal conceptualization of the individual, Maori people in the 1980s presented a potential risk to the legitimacy of the new vision because Maori aspirations were deeply located in history, in cultural differences and in the values of collectivity" (p. 249). She contends that the existence and practice of Maori visions, models, initiatives, and community struggles gave Maori a platform to challenge the neoliberal reform process. As she argues, "[i]t is important for communities to struggle, to act, to make sense of or theorise their experiences" (2006, p. 253).

Maori insistence that globalization be seen in a historical context also challenges what Jackson (2007a) called

> the jargon of globalised inevitability. . . . [T]he argument that there is no alternative or no other "reality", has become the seductive story that influences and changes our thinking about who we are and what we might become. In the genuine desire to improve the lot of our people, we lose sight of the fact that the New Right is the old righteousness of a colonizing order. Its apparent newness then makes its problems seem insoluble and its inequities seem intractable. (p. 172)

MAORI, INTELLECTUAL PROPERTY RIGHTS, AND TRADITIONAL KNOWLEDGE

Maori concerns about the WTO TRIPs (Trade-Related Aspects of Intellectual Property Rights Agreement) regime and threats to traditional knowledge have been widely expressed. With increasing pressures to harmonize intellectual property laws and growing commercial interest in indigenous knowledge, Maori knowledge and native flora and fauna have already been targeted by TNCs. One treaty claim (known as WAI 262) lodged over native flora, fauna, traditional knowledge, and intellectual property has enormous international significance. It is an assertion of sovereignty and directly challenges corporations that are commodifying and privatizing knowledge and biodiversity, helped by governments, which are revamping their patent laws for their benefit, and TRIPs. As Jackson dryly puts it, big business is eyeing a "whole range of "primitive newness" that has not yet been exploited in the current market environment" (interviewed in Maniapoto & Mills, 2005). He argues that Maori traditional knowledge cannot be squeezed into a colonial/Western legal intellectual property framework that denies spiritual, cultural, metaphysical aspects of traditional knowledge, as well as its collective nature.

> Because of an unsurprising absence of any notion of the sacred in the GATT concept of intellectual property, there is an obvious individualization and compartmentalization of knowledge based on its property value in commercial terms. The removal of knowledge from its sacred base in this way is the philosophical foundation which motivates Indigenous concern. Such concern is not based in any belief that knowledge possesses some quasi-Christian "soul" but rather upon the idea that knowledge and wisdom are part of the mauri [life essence] of existence. It is part of the interwoven threads of being and assumes its sacredness from the original *kete* [woven basket] of knowledge handed on to our ancestors. As such, it has traditionally been a collective entity, the use of which was regulated and protected. (Jackson, 1994, p. 4)

Linda Smith (1999) notes another tension inherent in the context of globalization and intellectual property rights: "The struggle for the validity of indigenous knowledges may no longer be over the *recognition* that indigenous peoples have ways of viewing the world which are unique, but over proving the authenticity of, and control over, our own forms of knowledge" (p. 104).

For Aroha Mead (Ngati Awa) (1993), misappropriation and commodification of indigenous knowledge represents

> [t]he new wave of colonization. . . . International agreements such as the GATT [now WTO] provide international acceptance for the principle of patenting all life forms, human as well as flora and fauna. . . . It can be referred to as "tampering", it can also be referred to as "misappropriation". Either way it is immoral and brings back painful memories of the attitudes of the first colonists, who regarded indigenous people as savages . . . What has changed? (pp. 7–8)

In 2002, Te Arawa/Tuwharetoa singer-songwriter Moana Maniapoto was threatened with a legal claim of 100,000 Euros by a German company should she sell CDs or perform there using her name, after it trademarked the word "Moana" for a range of products and uses (Maniapoto & Mills, 2005).

Maori women have been at the forefront of resistance, including their role in opposing GE and protecting traditional knowledge. Paul Reynolds (Tuwharetoa/Ngapuhi) and Cherryl Smith (Ngati Apa, Ngati Kahungunu, Te Aitanga a Hauiti) (2002) write about the significance of a national Maori women's network opposed to GE. Nga Wahine Tiaki o Te Ao collectively ran *hui* (meetings), which provided educational resources on GE. They note, "Whilst some groups have patronisingly claimed that Maori need more education on GE, Nga Wahine Tiaki has three women doctors, at least seven women with masters degrees, two lawyers and numerous other qualified people" (p. 41). Nga Wahine demanded that all GE material be removed from Aotearoa/New Zealand and for the cleaning up of GE-contaminated sites.

MAORI CRITICAL KNOWLEDGE, NGOS, AND ENVIRONMENTALISM

In settler-colonial states like Aotearoa/New Zealand, particularly for many NGOs based there, the dominant frame for most environmental, "global justice" or "antiglobalization" campaigns typically identifies TNCs, powerful governments like the United States, and domestic business and political elites as engines of neoliberalism, but essentially proposes a program of reforms and strengthening social democratic governance as a solution. This frame advocates nostalgia for a Keynesian welfare state, retooling the government, reregulating the economy, tighter controls on foreign investors, more social spending and more public consultation, and transparency around policymaking. Underpinning this formula are assumptions about supposedly universal and shared "Kiwi" values that must be reclaimed to (re)build a fairer society. I call this the "white progressive economic nationalist" position (Choudry, 2009, p. 99). It obscures and silences Maori histories of struggle for justice within and against the state. There is little reflexivity on the part of such NGOs about the knowledge on which they base their concepts of social justice and their own roles in reproducing colonial power relations. Largely missing from this dominant frame is any genuine acknowledgment of the colonial underpinnings of the state and society, the ongoing denial of Indigenous Peoples' rights to self-determination, and the highly racialized construction of New Zealand citizenship and state. For NGOs that address local issues, Maori are frequently reduced to a token sidebar in policy statements and declarations, a tragic case study, or otherwise rendered invisible or marginal in narratives designed to appeal to liberal audiences.

Several non-Maori environmental NGOs opposed the WAI 262 claim, notably the Royal Forest and Bird Protection Society, viewing themselves and the government as the rightful guardians of Aotearoa/New Zealand's biodiversity. For Jackson (1997), this NGO positioning is deeply colonial in itself:

> Conservationists have tended to adopt a narrow self-interested approach which in a revisiting of colonisation essentially claims that iwi have neither the right nor the ability to protect our resources. They seem to adopt the naïve view that the Crown should protect things for all New Zealanders, even though that is contrary to the Treaty, and even though the Crown is still attempting to sell of our assets to the highest multinational bidder. (p. 2)

Maori mobilization in defense of traditional knowledge and resistance to the imposition of intellectual property regimes offers further possibilities to popularize a decolonization position interwoven with a clear rejection of free market capitalism. Maori expertise and advocacy on the threats of intellectual property rights regimes has also been mobilized internationally in coalitions of other Indigenous Peoples. Meanwhile, Maori activism, education, and analysis have continued, including international networking, participation in People's Global Action (PGA) (Tuiono, 2007), Indigenous Peoples' networks and meetings at UN forums, and NGO/social movement forums contesting APEC and economic globalization (Jackson, 1999).

MEDIUMS OF CRITIQUE, LEARNING, AND CRITICAL KNOWLEDGE

During the 1980s and 1990s, through till today, Maori radio, hui, music, films, and other ways to disseminate analysis, increasingly the Internet, have been used. Cultural work, including music and film, has been an effective medium for Maori critiques of neoliberalism. In 1999, Tauni Sinclair (Ngati Porou) of Maori activist group Te Kawau Maro released the documentary film *Globalisation and Maori*. Maori resistance to colonization and imperialism expressed through music had a long lineage, and traditional forms mixed with diverse strands of contemporary music, not least, politically charged hip-hop/rap (e.g., recording artists such as Upper Hutt Posse, Te Kupu, Dam Native) and reggae (e.g., Aotearoa, Dread, Beat and Blood, Survival) in the 1980s and 1990s. Mauri Ora songwriter Takirirangi Smith's (Ngati Kahungunu/Ngati Apa/Aitanga a Hauiti/Ngai Tahu) *Tauiwi Ke* (Alien People) (T. Smith, 2001) denounces the appropriation and commodification of Indigenous Peoples' DNA by the Human Genome Diversity Project. After her disturbing experiences in Germany, singer/songwriter Moana Maniapoto filmed a documentary, *Guarding the Family Silver,* which critically examines the appropriation and commodification of Maori cultural icons and traditional knowledge for commercial use. During the 1990s, the International Research Institute for Maori and Indigenous Education (IRI) in the Maori Education department of the University of Auckland produced an excellent series of readers examining critical issues in contemporary Maori society with the theme "Economics, politics and colonization". Nga Kaiwhakamarama i Nga Ture/Maori Legal Service (of which Moana Jackson was director) produced important analytical resources encompassing both technical legal research and popular education approaches. Many other, often iwi-based or regional, education and awareness initiatives linked local issues with global and historical processes in ways relevant to Maori.

CONCLUSION

Cherryl Smith (2007) reminds us that

> [c]olonisation, like globalization, has inscribed various behaviours and ways of perceiving that go largely unquestioned in the world, both causing environmental and cultural destruction and posing solutions to them. They ignore issues as basic as understanding the importance of silence, of listening, of leaving certain areas untouched because they have stories and rights of their own, of respecting what belongs to others and of understanding that there is a place for continuity. Indigenous peoples continue to give voice to such simple and clear messages, but they still go unheard. (p. 73)

Maori resistance to domestic neoliberalism and global capitalism has continued to locate these processes in an understanding of colonialism. Bargh (2007a) documents Maori concerns and opposition to the WTO, MAI, and bilateral free trade agreements (see also Aotearoa Educators, 2000). 2004 saw massive opposition to legislation asserting Crown ownership of the country's foreshore and seabed, undermining Maori customary ownership. Similar moves were made on rivers and lakes, under the government's Sustainable Water Programme of Action (see Bargh, undated). In October 2007, a massive police operation locked down the Tuhoe Maori community of Ruatoki, with "antiterror" raids across the country, and arrests of mostly Maori activists amid sensationalist claims of guerrilla training camps in the bush. According to Jackson (2007b), "Maori see symmetries between the Terrorism Suppression Act and the 1863 Suppression of Rebellion Act. The targeting of mainly Maori as 'terrorists' in fact mirrors the earlier legislative labelling of those Iwi who resisted the land confiscations as 'rebels.'"

Moana Maniapoto's song *Te Apo* directly attacks the WTO, combining *haka* (dance/chant of challenge) and *karanga* (traditional call of welcome) with audio samples from the 2005 anti-WTO protests in Hong Kong. Its message is clear and uncompromising:

> *He kiri ki waho, he puku ki roto*
> Don't trust outward appearances, that which will only conceal what lies
>   within
> *Ana iā te apo, i ngā mahi tauhoko*
> *Homai rā tō ūpoko, 'merikana whanako.*
> Greed permeates the world of trade. Desist and apologize, American
>   plunderers
> *He kupu kei runga, he raku kei raro*
> Words float above, but deception lies beneath
> *Turakina te Roopu Tauhoko o te Motu*
> *taiao hurumutu, koia pū te utu.*
> The World Trade Organization needs to be stopped, or death to the environ-
>   ment is an inevitable.
> *Hīnana ki uta, hīnana ki tai*
> Keep the food sources of indigenous people protected, and well stocked—
>   these that are inherently theirs. (Maniapoto, Mills, Morrison, & Bridgman
>   Cooper, 2008).

Indigenous knowledge that explicitly links neoliberal capitalism and older forms of colonialism is often excluded, silenced, marginalized, or otherwise filtered to fit hegemonic positions dominated by NGOs and Northern activist networks. It is vital to retrieve the idea of self-determination as a conceptual tool of social transformation in an era in which it has been overwritten, or excluded from many scholarly analyses of imperialism and in the language of NGOs. If we are serious about achieving "global justice," we overlook such perspectives at our own risk. As Moana Jackson (2007a) puts it,

> There is no easy path to deconstruct globalization, to de-colonise the current forms of inequity and inequality, and there is an awful truth in Frederick Douglass's reminder power never gives up of itself without a struggle. It never has, and it never will, but knowing something of its whakapapa [genealogy] and finding power in our own stories and the alternatives they

might offer to change constitutions, economics and the ethics of life still offer hope for a better and more substantive enlightenment. (p. 182)

REFERENCES

Aotearoa Educators (2000, September 26). Prague-style protests to hit Aotearoa if Singapore deal continues. Press release. Retrieved from, www.scoop.co.nz/stories/PO0009/S00124.htm

Bargh, M. (ed.) (2007a). *Resistance: An indigenous response to neoliberalism.* Wellington: Huia.

————. A small issue of sovereignty. In M. Bargh (ed.), *Resistance: An indigenous response to neoliberalism* (pp. 133–146). Wellington: Huia.

————. (undated). Water under the bridge?. Retrieved from, http://www.conscious.maori.nz/news.php?extend.2

Burgmann, V., & A. Ure. (2004). Resistance to neoliberalism in Australia and Oceania. In F. Polet, & CETRI (eds.), *Globalizing resistance: The state of struggle* (pp. 52–67). London and Ann Arbor, MI.: Pluto.

Choudry, A. (2005, March). Crackdown. *New Internationalist*, pp. 16–17.

————. (2009). Challenging colonial amnesia in global justice activism. In Kapoor, D. (ed.). *Education, decolonization and development: Perspectives from Asia, Africa and the Americas* (pp. 95–110). Rotterdam: Sense.

Coney, S. (1997). *Into the fire: Writings on women, politics and New Zealand in the era of the New Right.* Birkenhead: Tandem Press.

Hyman, P. (1994). *Women and economics: A New Zealand feminist perspective.* Wellington: Bridget Williams Books.

Jackson, M. (1994, May). GATT and rangatiratanga: Tripping the lie fantastic. *Overview*, Corso, 50, pp. 1–2.

————. (1995, May). Maori lawyers express concern over foreign investment and treaty rights. *The Big Picture*, GATT Watchdog, 2, p. 3.

————. (1997, November). Flora, fauna, and the mysteries of GATT, APEC and MAI. *The Big Picture*, GATT Watchdog, 12, pp. 1–2.

————. (1999). Impact of globalization on marginalized societies and the strategies by indigenous people. In A. Tujan (ed.), *Alternatives to globalization: Proceedings of International Conference on Alternatives to Globalization* (pp. 101–106). Manila: IBON Books.

————. (2004). Colonization as myth-making: A case study in Aotearoa. In S. Greymorning (ed.), *A will to survive: Indigenous essays on the politics of culture, language and identity* (pp. 95–103). New York: McGraw-Hill.

————. (2007a). Globalisation and the colonizing state of mind. In M. Bargh (ed.), *Resistance: An indigenous response to neoliberalism* (pp. 167–182). Wellington: Huia.

————. (2007b). Back in the mists of fear. A primer on the allegations of terrorism made during the week October 15–17. Retrieved from, http://www.converge.org.nz/pma/mj231007.pdf

Jesson, B. (1999). *Only their purpose is mad.* Palmerston North: Dunmore Press.

Joseph Rowntree Foundation (1995). *Income and wealth: A report of the JRF Inquiry Group.* Retrieved from, http://www.jrf.org.uk/publications/income-and-wealth-report-jrf-inquiry-group

Kelsey, J. (1995). *The New Zealand experiment: A world model for structural adjustment?* Wellington: Bridget Williams Books.

———. (1999). *Reclaiming the future: New Zealand and the global economy.* Wellington: Bridget Williams Books.

———. (2002). *At the crossroads—Three essays.* Wellington: Bridget Williams Books.

Maniapoto, M., & T. Mills. (dir.) (2005). *Guarding the family silver: Intellectual property and cultural appropriation.* Auckland: Tawera/Black Pearl Productions.

Maniapoto, M., T. Mills, S. Morrison, & M. Bridgman Cooper. (2008). *Te Apo.* From the music CD *Wha.* Black Pearl/Ode Records. Retrieved from, http://moananz.com/popups/lyrics/te_apo.html

Mead, A. (1993). Misappropriation of indigenous knowledge: The next wave of colonization. Retrieved from, http://www.kaupapamaori.com/assets//MeadA/nga_tikanga_nga_taonga_misappropriation_of_indigenous_knowledge.pdf

National Maori Congress (1994, November 29). Press release.

Pihama, L. (1999, November). APEC: Colonial manipulations. *The Big Picture,* GATT Watchdog, 19, pp. 4–6.

Reynolds, P., & C. Smith. (2002). *Aue! Genes and Genetics.* Whanganui: Whanganui Law Centre.

Rosenberg, W. (1993). *New Zealand can be different and better: Why deregulation does not work.* Christchurch: New Zealand Monthly Review Society.

Sinclair, T. (dir). (1999). *Globalisation and Maori.* Auckland: Te Kawau Maro.

Smith, C. (2007). Cultures of Collecting. In M. Bargh (ed.), *Resistance: An indigenous response to neoliberalism* (pp. 65–74). Wellington: Huia.

Smith, G. H. (1993, November). The commodification of knowledge and culture. *Overview,* Corso, 49, pp. 6–7.

———. (1998). Iwi wars: Neo-colonisation in Aotearoa. In International Research Institute for Maori and Indigenous Education, *Fisheries and commodifying iwi: Economics, Politics and Colonisation, 3* (pp. 48–51). Auckland: IRI/Moko Productions.

Smith, L. T. (1999). *Decolonising methodologies: Research and indigenous peoples.* London and New York: Zed Books.

———. (2006). Fourteen lessons of resistance to exclusion: Learning from the Maori experience in New Zealand over the last two decades of neo-liberal reform. In Mulholland, M. and contributors, *State of the Maori nation: Twenty-first century issues* (pp. 257–249). Auckland: Reed.

Smith, M. (1998). The fisheries settlement in the context of wider Maori development issues. In International Research Institute for Maori and Indigenous Education, *Fisheries and commodifying iwi: Economics, Politics and Colonisation, 3* (pp. 43–46). Auckland: IRI/Moko Productions.

Smith, T. (2001). Tauiwi Ke. From *Ki te ao marama* (music CD) by Mauri Ora. Wellington: Whenua Records.

Tuiono, T. (2007). We are everywhere: Interview with Teanau Tuiono. In M. Bargh (ed.), *Resistance: An indigenous response to neoliberalism* (pp. 125–129). Wellington: Huia.

UNCTAD (2000). *World investment report. Cross-border mergers and acquisitions and development.* New York and Geneva: United Nations. Retrieved from, http://www.unctad.org/en/docs/wir2000_en.pdf

Walker, R. (1990). *Ka whawhai tonu matou: Struggle without end.* Auckland: Penguin.

CHAPTER 5

ETHNIC MINORITIES, INDIGENOUS KNOWLEDGE, AND LIVELIHOODS: STRUGGLE FOR SURVIVAL IN SOUTHEASTERN BANGLADESH

BIJOY P. BARUA

INTRODUCTION

Indigenous knowledge is neither static nor frozen. Rather it is socially dynamic and culturally appropriate as it is practiced in the local environment. It has evolved from years of collective learning experience through the process of trial and error. Indigenous knowledge confronts the commodification of science and the consumer paradigm. This knowledge is developed through a practice of experiential learning by the people within the geographic area or culture. It is derived from multiple sources within the natural environment; it is holistic and inclusive (Barua & Wilson, 2005; Dei, Hall, & Rosenberg, 2000). Such indigenous knowledge has been neglected or stripped down in the implementation of the Western development model in rural Bangladesh. Over the years, the Western model of development has disregarded the diversity of indigenous knowledge and culture in order to promote a Eurocentric-monoculture in the southeastern part of

Bangladesh. The enormous diffusion of commercialization tends to regulate ethnic minorities' (i.e., the rural *Chakma* and *Marma* communities) ways of life through centralized urban control and the construction of a dam[1] and establishment of various projects for industrialization. Over the last four decades, the oppressive politics and policies of development have eroded the decentralized decision-making process, knowledge, culture, and livelihoods of ethnic minorities in the hilly land. However, despite massive expansion of the Western model of development and sociocultural oppression, there is a growing cultural and political resistance among the ethnic minorities to regenerate local knowledge and livelihoods for cultural rights in rural Bangladesh (Barua, 2004, 2007; Barua & Wilson, 2005).

This chapter advances a critique of Western assumptions about development that push indigenous knowledge, culture, and livelihoods to the margins in the Chittagong Hill Tracts of Bangladesh. The chapter will be based on my experience and research[2] in the areas of Kaptai and Rangamati (Barua & Wilson, 2005). In the chapter, I also critically examine the sociocultural resistance movement of ethnic minorities of the Chittagong Hill Tracts within the framework of indigenous knowledge, culture, development, and movements. While critiquing the notion of development programs in the context of the Hill Districts, I review the literature on indigenous knowledge, development, and livelihoods in order to construct a conceptual framework for interpretation. In this chapter, I specifically focus on the issues of ethnic minorities, colonial policies, forests and livelihoods, development interventions, role of development actors, and minority struggle and resistance. For convenience of discussion, I will use the expressions "ethnic minorities," "ethnic communities," and "hill people" interchangeably.

INDIGENOUS KNOWLEDGE, DEVELOPMENT, AND LIVELIHOODS

The failure of the development approach and model in developing countries has become a concern for the policy planners and development thinkers for the last four decades in the international forum. Academic interest in "indigenous knowledge" of developing countries has evolved since the publication of the book, *Indigenous Knowledge System and Development* by Brokensha, Warren, and Warner in 1980 (Kothari, 1996). This book explicitly raised critical questions about the production of knowledge in the area of development and rural people in developing countries. In other words, this publication had also challenged dominant Western knowledge in the area of development. As a result, many social and natural scientists came forward to acknowledge the contribution of indigenous knowledge for development. This has eventually made the issue of indigenous knowledge systems a part of culture, which has served to create a conducive environment for participatory approaches in the field of development. Eventually, the proclamation of the World Decade for Cultural Development (WDCD) in the mid-1980s, the Social Summit in Copenhagen in 1990, and the Earth Summit of Rio in 1995, have also increased awareness among international development agencies to adopt the issue of indigenous culture, knowledge, and sustainable agenda through a participatory approach within their program policies (United Nations, 1990; UNESCO, 1995).

Indigenous knowledge is nonformal education and close to the nature. In addition, "[i]ndigenous knowledge(s) differ(s) from conventional knowledge(s) because of an absence of colonial and imperial imposition" (Dei, Hall, & Rosenberg, 2000, p. 7). It is learned neither in a formal educational environment nor in isolation. Rather indigenous knowledge is gained from the natural environment through communal activity. It derives from multiple sources within the natural environment (Dei, Hall, & Rosenberg, 2000).

Indigenous knowledge systems have been prevalent in societies for a long time. They are experiential and tend to address diverse and complex conditions of the society and environment for sustainable livelihoods. However, indigenous knowledge systems have drifted to the margin as a result of the influence of Western knowledge and development. Indigenous knowledge is maintained and transmitted through an oral record and retained over many centuries within a specific geographical area or culture for the sustainable development of that community (Debrah, 1994; Deshler, 1996). Dei, Hall, and Rosenberg (2000) conceptualized indigenous knowledge as "a body of knowledge associated with the long-term occupancy of a certain place" (p. 6) and as refined through a process of experimental learning over the years in order to address a specific natural context and environment for sustainable growth. Warren (1991) defined indigenous knowledge as

local knowledge—knowledge that is unique to a given culture or society— IK contrasts with the international system generated by universities, research institutions and private firms. It is the basis for local level decision making in agriculture, health care, food preparation, education, etc. (p. 1)

While generating indigenous knowledge, tangible resources such as, land and cattle, sociocultural environment, and historical context are also considered as key factors of livelihoods.[3] Indigenous knowledge encourages a participatory decision-making process for the purpose of effective development of local organizations and people and to promote biodiversity for food and medicine (Shiva, 2000a). For ethnic communities, "conserving biodiversity means conserving their rights to their resources and knowledge, and to their production systems based on biodiversity" (Shiva, 2000b, p. 28). Despite its richness, this knowledge is "underutilized" (Howes & Chambers, 1980, p. 329) in the grassroots development in developing countries. In fact, indigenous knowledge systems help develop a self-reliance capacity rather than dependency for livelihoods within the community (Shiva, 2000a). Undeniably, this "indigenous knowledge is a part of an integrated whole" (Brokensha, Warren, & Werner, 1980, p. 4) among the rural people. If development planners fail to design programs based on the local context, they are naturally geared toward failure rather than success. For example, several development projects such as the Narmada Dam in India, the James Bay in Canada, the Three George Dam in China, and the rubber cultivation project and Eco-Park (funded by the Asian Development Bank) in Modhupur forest in Bangladesh have failed to address the local knowledge, environments, and livelihoods of the indigenous people and peasant communities (Gain, 1998; Miles, 2004; Shiva, 1989).

While developing projects for the development of the community, it is essential to address the issues of local needs, local knowledge, resources, and spirituality rather than the material aspect of programs for the development of people. Indeed, "[k]nowledge gained through spiritual means can serve economic as well as psychological needs" (Castellano, 2000, p. 24). Despite this fact, the development model has treated "indigenous knowledge as inefficient, inferior, and an obstacle to development" (Agrawal, 1995, p. 403). While claiming this, indigenous knowledge is portrayed as backward and irrational in order to modernize the indigenous people and rural people through the growth model in developing countries. On the other hand, advocates of indigenous knowledge are critical of the Western development model as this involves commodification and commercialization of knowledge in the name of scientific development. While critiquing, the theorists of indigenous knowledge view Western science as nonresponsive to local desire and demand and as irrelevant and questionable to the local community. Nevertheless, indigenous knowledge is highly varied and location-specific (Brokensha, Warren, & Werner, 1980). Unquestionably, these two forms of knowledge use different methods and processes to explore reality due to the differences of epistemological grounds. In reality, "no knowledge system can exist in a cultural, economic or political vacuum" (Semali, 1996, p. 13). Indisputably, indigenous knowledge procedures are appropriate for both forests and communities. The traditional values and local knowledge and wisdom are considered to be the key formation of sustainable development and organizations within indigenous communities. Indigenous knowledge of forest science does not perceive trees as merely wood and market commodity. Rather, it focuses on diversity of form and function for the survival of humans and the prolongation of natural forests (Shiva, 1989). While considering this agenda of indigenous knowledge, the Intergovernmental Panel on Forests (IPF) adopted element 1.3 in its program for the sustainable development of forests:

> Consistent with the terms of Convention on Biological Diversity, encourages countries to consider ways and means for the effective protection and use of traditional forest-related knowledge, innovations and practices of forest-dwellers,

Indigenous people and local communities, as well as fair and equitable sharing of benefits arising from such knowledge, innovations and practices.

cited in Battiste, 2000, p. 260

However, the adaptation of indigenous knowledge within the mainstream development programs by the international development agencies is becoming marginalized due to the technocratic approach. Moreover, the development intervention toward empowerment and liberation tends to be appropriated by the urban experts with acceptance of participatory approach in developing countries in recent times (Rahman, 1995). Despite this, the theory on development must be of practical relevance to the communities in order to maximize the benefit for the people (Ariyaratne, 1996).

ETHNIC MINORITIES AND THE SOCIO-POLITICAL AND CULTURAL CONTEXT

Ethnic minorities mainly live in the regions of northeastern, south eastern, central eastern, and northwestern parts of Bangladesh. It is estimated that a total of 12,05,978 ethnic minorities live in the country. Ethnic minorities are also known as *adivasis* in Bangladesh. These groups are culturally distinct from one another and a total of forty-five ethnic minority groups live in the country (Barua, 2004). Among these groups, the *Chakma* and *Marma* communities live in the hills of the southeastern part of Bangladesh, mostly in the Hill Districts of Bandarban, Kahagrachari, and Rangamati. The socio-political and cultural foundations of these communities are deeply rooted in the organization of kinship ties in the Hill Districts. In other words, this lineage based indigenous organization is often used in order to serve the interest of colonial administration. In many instances, the chieftains of the Hill regions were used as puppets to collect revenues and to control the administration (Dewan, 1993).

The *Chakma* community is socially divided into forty kinship groups and the *Marma* are divided into thirty kin groups. Over the years, the *Chakmas* have adopted Bengali personal names. On the other hand, the *Marmas* maintain the *Arakanese* personal names as part of their cultural identity. Their physiognomy is fairly different from the mainstream population of Bangladesh. These communities are also known as *Magh*[4] in the Hill Districts and Chittagong. These communities practice the *Theravada* tradition of Buddhism (Barua, 2007). Although the political structures of the valley-living *Chakma* and *Marma* have a certain commonality, they are quite different from other ethnic minorities of the hills. The *Chakmas* speak a colloquial form of Bengali, which is more similar to *Ahomian* and *Chittagonian* languages and belongs to the Indo-Aryan language group. On the other hand, the *Marmas* speak *Arakanese*, which is written in Burmese characters (Barua, 2001). The *Chakma* and *Marma* communities maintain their own distinct sociocultural traditions in the hills of Bangladesh. Although these communities maintain distinct cultural identity through the symbol and essence of *jhum* (slash and burn/swidden agriculture) cultivation, the *Chakmas* acculturated and assimilated "to the ways of life and modes of thought of the *Bengali* communities with whom they came in contact" (Bertocci, 1984, pp. 349–350) over many years. Their elites are highly *"Hinduized"* (Bertocci, 1984, p. 347) in the urban hubs. Traditionally, the mainstay of the economies of these communities was agriculture, primarily *jhum* and plough cultivation in the valleys (Dewan, 1990). These communities remain exclusively dependent on the resources available in the forests of their hilly environment for their livelihoods (Rahman, Khisa, Uddin, & Wilcock, 2000). The cash crop was introduced in the region during the periods of the Mughal (1666–1760) and the British (1760–1860) for acceleration of economic development. Despite this, food insecurity is massive in the Hill Districts. The ethnic communities of the hills are in a desperate situation when collecting food during *Ashar* (June–July) and *Sravan* (July–August) each year despite the introduction of a growth-based development model in the region. About 62 percent of households, regardless of ethnic groups and identities, are living below the absolute poverty line (below 2,122 kilo calories) in the region (Barkat, 2009).

COLONIAL POLICIES, FORESTS, AND LIVELIHOODS OF ETHNIC COMMUNITIES

Forest economy requires balancing farming with the use of natural forests in order to maintain natural fertility of the soil and to promote sustainable development for ethnic communities. In other words, forest economy encourages the growth of natural plants and helps raise livestock that meet the basic needs of the forest communities. In this effort, the forests and hill communities are eco-friendly and tend to maintain harmony with the natural environment and preserve local plants and traditions in their environment. The living nature is the source of knowledge and livelihoods. The continued existence of the forests is necessary for the survival of *dhamma* (education and morality) and the environment (Barua & Wilson, 2005). Moreover, it allows "for sustainable food production systems in the form of nutrients and water" (Shiva, 1989, p. 65).

The ownership of lands and forests among the ethnic communities in the hills is not private. Rather, lands and forests are managed and controlled by the village community through collective efforts (Zaman, 1984). However, the British-Indian Forest Act of 1865 declared forests and wastelands as reserved forests in order to restrict and control the communities. Although this was adopted in the name of scientific management of forests and lands, the forest policies and programs have had a direct effect on the destruction of the forests, largely displacing ethnic communities, disenfranchising them of their livelihoods, and dislocating them from the locus of their sociocultural beliefs and practices in the Hill Districts. Bangladesh inherited the policies and regulations from the British colonial authority, which allowed the exploitation of the forests and natural resources for economic profit and benefit (Barua & Wilson, 2005). In other words, the colonial acts and rules had initiated the process of detribalization of tribal lands and forests through encroachments on their own terrains (Kapoor, 2007). Such policies of land and forest encroachments have failed to address the socioecological and cultural values of natural resources. Moreover, the forest acts and policies have tended to ignore the customary rights and indigenous knowledge of the ethnic communities. Over the period, many of these policies have remained in effect during the rule of erstwhile Pakistan and Bangladesh. Furthermore, the successive governments have integrated many rules and regulations to control the forests and natural resources in the region (Barua & Wilson, 2005). Mohsin (2003) stated critically:

> Successive regimes have alienated the hill people and shown no consideration for their traditional customary rights, which are protected under Regulation 1 of the 1900 CHT manual (currently in effect). The government has provided neither alternative sustainable means of employment nor compensation for lands and resources appropriated. (p. 27)

In the last four decades, the Forest Acts have been modified and reoriented to allow the spread of industrialization and modernization into the hill areas, largely at the expense of forestlands and natural resources. The first extensive commercial teak plantations were initiated in 1871. As a result, plant life has turned into a nonrenewable resource. Diverse varieties of trees have been replaced by monocultivation in the forest. Between 1919 and 1923, plantations spread throughout the region (Mohammud, 2005). Successive development policies permitted the planting of massive plantations and mega projects accelerated the destruction of ecosystems and biodiversity. Particularly, schemes for rubber production accelerated and it had overtaken large portions of the arable land in the Hill Districts by 1980. The expansion of reserved forest and commercial plantation projects has not only restricted community access to the forest, but it has also driven out the hill people of their livelihoods practices such as *jhum* cultivation, horticulture production, raising livestock, and natural gardening (Rafi, 2001). Amar of the *Chakma* community expressed,

> People used to raise and nurture litchi, jackfruit, mango and guava gardens in our villages. Cow, sheep and cattle were also essential domestic animals for livelihood of the people. Nowadays, people are not able to maintain these domestic animals due to lack of grazing lands. Over the period, the economic condition has changed drastically. Even, these days, people neither raising gardens nor practicing *jum* cultivation.
>
> Personal Communication, 2009

By 1992, some 4000 hectares of forest were given over to rubber plantations (Mohsin, 1997). Commercial rubber and eucalyptus plantations encroach and impose on community-based gardening in the forest, resulting in loss of biodiversity, community access, and self-sufficient community economy. The loss of local varieties of trees and forests has further aggravated the decline in Buddhist spiritual values and ethical practice in villages (Barua & Wilson, 2005). The commercial plantation schemes also wiped out the native herbs and wild vegetables, which are part and parcel of the lives of the forest and the hill people.

The establishment of the Karnaphuli Paper Mill[5] in the Hill Districts in 1953 exemplifies the process by which modernization and industrialization are resulting in deforestation and displacement of the *Marma* community (Gain, 2000). The paper mill is one of the major sources of deforestation in the hilly land. Since the establishment of the paper mill, millions of tons of bamboo and softwood have been used for paper production (Rafi, 2001). In particular, *Marma* communities were displaced from the area of the mill in order to permit access to the forests and other natural resources. In other words, the mega project displaced the community from their livelihoods. While doing so, the economic interest of ethnic community, particularly women were ignored, who depended on the natural resources of the forests for their survival. In fact, the mill has created more job opportunities for nonethnic communities who occupy all major positions in it, while only about 1 percent of ethnic minorities are employed at the lower levels. Such industrialization brought no benefit in terms of employment to the ethnic *Marma* minorities in the Hill Districts (Rafi, 2001). Rather, it destroyed the local culture, economy, and environment for the sake of commercialization of forest products (Barua &Wilson, 2005). The living forest was their shelter, food, and their source of light and water (Shiva, 1989). Aung, a member of the *Marma* community explicitly mentioned:

> In the past, villagers lead a simple life through the *jhum* cultivation (swidden agriculture) and plough cultivation. Our economy was deeply rooted in the forest and land. Unfortunately, the state authority restricted our access to the forest and land. Hence, our livelihoods have been dislocated. Money economy has become dominant. Now people are struggling to survive in our lands.
>
> Personal Communication, 2009

In the last four decades, the scheme of industrialization and modernization has not only displaced the rural *Chakma* and *Marma* communities from their own livelihoods, it has also diminished their spiritual and ethical values. Moreover, for the people of Hill Districts, their land is in a sense like their mother. They are respectful of their land since it provides food and shelter to them. The *Chakma* and the *Marma* communities maintain a spiritual connection with the land and forest through the diverse cultural festivals and rituals. Often such rituals and cultural festivals are organized through *halpalni* (worship of land and forest) of the *Chakma* community and *row sangma puja* (worship of trees and forest) of the *Marma* community (Chakma, 2007; Shoenu, 2007) to pay respect to the lands and forests. For example, these cultural festivals are mainly offered in the celebration of *puja* (worship) and *dana* (gift/sharing merits) to the land. In this environment and process, no project will be successful and sustainable if the economic context of the people is disjointed from the sociocultural and spiritual aspects of the people (Ariyaratne & Macy, 1992). The loss of cultural and spiritual values and the local economy have disintegrated the communal harmony in the name of modernization and urbanization in the hilly lands. Over the years, many Buddhist ideals and practices have been disappearing due to the commercialization of natural forest, such as the *dhutanga* (ascetic) practices of monks that provide the basis for contemplative education and the perpetuation of traditional perspectives and knowledge. Eventually, the ethnic communities became poorer as they lost their control over the natural resources. A similar disappearance was also observed in Northeastern Thailand (Barua & Wilson, 2005).

DEVELOPMENT, DAMS, AND EDUCATION: CULTURAL RESISTANCE AND POLITICAL STRUGGLE

The present development model and intervention have failed to provide local specific knowledge to the rural *Chakma* and the *Marma* communities based on indigenous knowledge, right livelihoods, and Buddhist values and ethics in the region of Hill Districts. More importantly, commercialization of forestland for cash crop (i.e., tobacco production) and the introduction of the Western development model for economic growth have displaced the socioeconomic and cultural fabrics of the people in the hilly land. Over the years, development projects have trigged violence in the hills. Such structural violence demobilized the rural communities and destroyed their local self-sufficient economy. The construction of the *Kaptai* hydroelectric dam[6] in the 1960s resulted in disaster for the *Chakma* community (Dewan, 1990; Zaman, 1996), including the inundation of some 10 square miles of reserved forest, 54,000 acres of arable land, and the displacement of approximately 100,000 environmental refugees, which was one-sixth of the ethnic communities of the hills (Samad, 1994, 2000). Mohsin and Ahmed (1996) more explicitly expressed:

> The construction of the dam had far-reaching consequences for the tribal (ethnic communities) people . . . It made nearly 10,000 *Chakma* ploughing families having proprietary rights, and 8,000 *Chakma jhumia* families comprising more than 10,000 *Chakma* persons are land less and homeless. It also affected . . . 1,000 *Marmas*. (p. 279)

This mega development has not only marginalized the ethnic communities and created dependency on the external inputs of the so-called experts in the Hill Districts, but it has also shown disrespect to the land and nature. It also

dislocated ethnic communities from their natural resources and livelihoods. The growth-oriented model of development emphasizes a capital,
technology, and energy intensive approach. This model is confined to the
matters of production, land control, cash, and profit at the cost of human
lives and local economies (Barua, 2009). As a result, the people lost access
to common property resources, right of entry to grazing lands, and cattle.
In other words, the mega project was built in the Hill District in order to
industrialize the subsistence economy of the country for economic growth
within the framework of a modernization model and approach. Although the
dam was constructed for higher economic growth, it created serious threats
to the livelihoods of the rural ethnic communities in the hills. As Shiva
(1989) writes, this

> culturally biased project destroys wholesome and sustainable lifestyles and cre-
> ates real material poverty, or misery, by the denial of survival needs themselves,
> through the diversion of resources to intensive commodity production. (p. 10)

Eventually, the subsistence economy of the ethnic communities was transformed into industrialization and urbanization at the cost of their lives and
livelihoods (Dewan, 1993). As a result, the construction of the *Kaptai* Dam
neither ensured livelihood sustainability to the rural *Chakma* and *Marma*
communities nor maintained ecological sustainability in the region in the
name of development. In other words, Western science attempts to limit its
quest for a universal knowledge of nature in the name of progress and development without acknowledging the people's science and knowledge. On other
hand, the primary quest of indigenous science and knowledge is to treat the
nature of human beings as an inseparable part of nature itself for sustainable development (Ariyaratne, 1996). Despite these facts, Western science
has been engaged in developing other societies for the purpose of industrial
exploitation in the name of growth and progress. More often, such development projects have tended to take nonparticipatory approaches with the
financial assistance of international agencies in developing countries.

Mainstream development planners and implementers have tended to
change the sociocultural values, knowledge, and appropriate technologies of
the people through processes of nonformal and formal education. In many
instances, such educational models focused on high technology, urban-life
pattern, and cities without embracing location-specific knowledge and ethical
issues (Norberg-Hodge, 1991). This education also worked hand in hand
with consumerism in order to make planetary crisis through the path of
market economy (O'Sullivan, n.d.). Over the years, both formal and nonformal education was designed to desensitize the ethnic communities in the
name of industrialization and urbanization. It also dehumanized and denigrated
local cultures and environments in the name of progress and scientific development. For instance, the *Chakmas* of the hills living in the urban areas embraced

the Western form of education due to the change of natural environment and economic conditions (Barua, 2007, 2009). Despite displacement of these communities, the development programs neither rehabilitated nor addressed the needs and demands of the rural ethnic communities in their homeland. Although the Chittagong Hill Tracts Development Board (CHTDB) was created in 1976 to implement the massive development projects in the hills with the financial assistance of international donors, it failed to address the interests of the rural ethnic communities.[7] Moreover, the government also rehabilitated nonethnic communities from the plain land into lands of ethnic communities. In many cases, the rehabilitation projects favored mainland settlers (Dewan, 1993; Mohsin, 1997). While rehabilitating the nonethnic groups in the lands of ethnic communities, the development agencies and departments unfortunately ignored the rehabilitation programs for the ethnic communities who were displaced due to the construction of the dam in the hills. Amar said,

> We became refugees in our own land. Our cultivable lands were submerged under water. We do not have access to the forest for *jhum* cultivation. Moreover, the outside people were rehabilitated in our lands. We constantly encounter problems from the settlers. We are isolated from our lands. (Personal Communication, 2009)

Such rehabilitation programs for the nonethnic communities created more precarious conditions for the Hill people and instigated problems of land disputes in the area (Mohsin, 1997). Consequently, this top-down model of development initiatives toward community empowerment and rehabilitation schemes practically initiated the process of "indignation and bitterness rather than mitigating grievances" (Haque, 1990, p. 51) among the ethnic communities. As a result, political resistance against social injustice and oppression was mobilized in the hills by the ethnic communities through political movements in the early 1970s.

In mobilizing the Hill people, schoolteachers and educated youth groups were actively involved in the resistance movement through the formation of *Parbattya Chattagram Jana Sanghati Samity* (PCJSS: United People's Party of Chittagong Hill Tracts). Eventually, the PCJSS organized the *Gono Mukti Fouj* (People's Liberation Army), generally known as the *Shanti Bahini* (Peace Corps) in the country (Haque, 1990). Many of these groups were motivated and influenced by socialist ideology in fighting to create equitable society for the ethnic communities. Over the years, the political resistance further intensified through armed struggle and insurgency movements based on Marxist-Leninist ideology (Barua, 2001; Zaman, 1984). In course of time, the sociopolitical struggle turned into an indigenous social movement aiming to establish *jumma* (the distinctiveness of the ethnic communities who reside in the hills) land for the ethnic communities. However, the members of other ethnic communities of the Hill, particularly, the *Marmas* and

Tripuras rejected the notion of *jumma* nationalism as the word endorses the hegemony of the *Chakma* community (Tripura, 2000). There is also great debate among the educated members of the *Chakma* community about the appropriateness of the term *jumma* as the word is often used in a derogatory sense in the country. While embracing the notion of indigenous social movement through the construction of *jumma* nationalism in the hills by the political activists, it tended to ignore the spirituality of the people. Despite the substantial sociopolitical and cultural resistance against colonialism, development agencies, both national and international, have been engaged in natural resource extraction and economic exploitation in the region under the scheme of modernization and development (Dewan, 1993).

On the other hand, the sociopolitical movement of the Hill Districts practically created new urban political elites and opportunist classes among the ethnic communities and dominated by the members of the *Chakma* community. In this situation, the *Marmas,* who mostly live in the rural areas, are facing difficult conditions and having troubles establishing their political rights in the hills. Although the Peace Accord was signed in 1997 to create social stability and sustainable development in the region, it could not address the demands and needs of the ethnic communities (Mohsin, 2003). Despite this, "the accord has also established the hegemony of the Chakma, who are numerical majority and the dominant community" (Mohsin, 2003, p. 53) in the hills. However, the Peace Accord has paved the path for the international development agencies and national nongovernmental organizations (NGOs) to improve the standard of living for the people of the hills. The people have increasingly witnessed the emergence of NGOs in the post-accord period in the Hill Districts. There has been a growing trend among the educated urban ethnic communities to establish local NGOs in order to initiate development programs for the rural ethnic communities within the hills,[8] a total of fifty-two local non-governmental organizations (LNGOs) have emerged in the hills (Mohsin, 2003). These local NGOs have been working with the financial support of bilateral and international organizations in the areas of income generation, social forestry, horticulture, fisheries, poultry farming, health, water, and sanitation.[9] Because of this, the local NGOs tend to adopt the approach of service-oriented programs without any analytical and critical thinking. On the other hand, the national NGOs have been preoccupied with microcredit programs in the hills, targeting disadvantaged women in particular. These national NGOs are mostly working around cities. Their projects tend to undermine the sociocultural values, knowledge, and aspirations of the hill people (Mohsin, 2003). Such a model explicitly follows the trickle down approach instead of a bottom up approach. Despite this, there has been resistance from the ethnic communities against national NGOs that are actively engaged in implementing microcredit programs with high interest in order to transform the subsistence economy and culture in the name of growth and progress. Although a few national NGOs have been engaged in developing local languages in the education program for the appropriation of foreign funding, the ethnic minorities are dissatisfied with the nature of their project activities. In such a context, both the national NGOs and local NGOs have become competitive with each other in protecting and strengthening their own financial resources instead of focusing on people-centered development. Inevitably, the people-centered development embodies the principles of social justice, equity, peace, sustainability, and inclusiveness. It acknowledges the peoples' knowledge, local economics, and ownership (Korten, 1990).

CONCLUDING REFLECTIONS

In this chapter, I have critically examined the issues of development models, trends, and livelihoods of ethnic communities within the framework of indigenous knowledge, culture, politics, and movement in the context of Hill Districts of Bangladesh. I have shown that the development policies and programs in the hills have followed the path of the colonial legacy that promoted centralized control through the process of imposition and domination. Although the ethnic minorities of the hills have always been impassioned and relentless about establishing their cultural rights and identities through a political struggle for their livelihoods and economic survival in southeastern Bangladesh, they have been trapped in the vicious circle of colonialism through the appreciation of a Western model of development. Unfortunately, such a model has failed to recognize the psychological, social, cultural, and spiritual aspects of the ethnic communities in the development process. In other words, this development model disregarded the indigenous knowledge, culture, and natural environment. However, interestingly, the Peace Accord laid the path for international development agencies, bilateral and multilateral donors, and national NGOs to invest funds and materials for implementing development projects in the hills. Simultaneously, this accord also created a place for the educated ethnic groups of the urban hubs to establish LNGOs in the hills. Importantly, those who established these LNGOs did not have adequate exposure to the development processes and interventions. Because of this, the development actors are neither promoting people-centered development nor nurturing indigenous knowledge for sustainable livelihoods of the rural ethnic communities in the hills and forests. Rather, all these NGOs tend to promote a trickle-down approach to economic growth and progress in the hills of southeastern Bangladesh. Such an approach will certainly engender more competition and confrontation among the various actors and local communities, albeit in the name of peace and development.

REFERENCES

Agrawal, A. (1995). Dismantling the divide between Indigenous and scientific knowledge. *Development and Change, 26*, 413–439.

Ariyaratne, A. T. (1996). Village studies for development purposes (pp. 77–86). In N. Ratnapala (ed.), *Buddhism and sarvodaya, Sri Lankan Experience*. Delhi: Sri Satguru Publications.

Ariyaratne, A. T., & J. Macy. (1992). The island of temple and tank, sarvodaya: Self-help in Sri Lanka (pp. 78–86). In M. Batchelor & K. Brown (eds.), *Buddhism and ecology*. London, England: Cassell Publishers.

Barkat, A. (2009). *Life and livelihood of indigenous peoples in the Chittagong hill tracts: What we know?* Paper presented at the Regional Seminar, Organized by Bangladesh Economic Association Chittagong Chapter and Economic Department, February 28, 2009, University of Chittagong.

Barua, B. P. (2001). *Ethnicity and national integration in Bangladesh: A study of the Chittagong hill tracts*. New Delhi: Har-Anand Publications.

——— (2004). *Western education and modernization in a Buddhist village of Bangladesh: A case of the Barua community*. Unpublished doctoral dissertation, Department of Sociology and Equity Studies in Education, Ontario Institute for Studies in Education, University of Toronto.

——— (2007). Colonialism, education and rural Buddhist communities in Bangladesh. *International Education, 37*(1), 60–76.

Barua, B. (2009). Nonformal education, economic growth and development: Challenges for rural Buddhists in Bangladesh (pp. 125–140). In A. Abdi & D. Kapoor (eds.), *Global perspective on adult education*. New York: Palgrave Macmillan.

Barua, B., & M. Wilson. (2005). Agroforestry and development: Displacement of Buddhist values in Bangladesh. *Canadian Journal of Development Studies, 26*(2), 233–246.

Battiste, M. (2000). Reclaiming Indigenous voice and vision. Vancouver, Canada: UBC Press.

Bertocci, P. (1984). Resource development and ethnic conflict: The case of Chittagong Hill Tracts of Bangladesh (pp. 345–361). In M. S. Qureshi (ed.), *Tribal culture in Bangladesh*. Rajshahi, Bangladesh: Institute of Bangladesh Studies, Rajshahi University Press.

Brokensha, D. W., D. M. Warren, & O. Werner (1980). Introduction (pp. 1–8). In D. W. Brokensha, D. M. Warren, & O. Werner (eds.), *Indigenous knowledge systems and development*. Lanham, MD: University Press of America.

Carino, J. (1999). The world commission on dams: A review of hydroelectric projects and the impact on indigenous peoples and ethnic minorities. *Cultural Survival Quarterly, 23*(3), 53–56.

Castellano, M. B. (2000). Updating aboriginal traditions of knowledge (pp. 21–36). In G. S Dei, B. L. Hall, & D. Goldin-Rosenberg (eds.), *Indigenous knowledges in global contexts: Multiple readings of our world*. Toronto, Ontario, Canada: University of Toronto Press.

Chakma, S. (2007). Chakma Utshob O Bibah [Festivals and marriage of the Chakma community]. In S. Chakma (ed.), *Parbarto Chotrogram Upojati Uthshob O Bibah*. [Tribal festivals and marriage of the Chittagong Hill Tracts]. Rangamati, Bangladesh: Tribal Cultural Institute.

Debrah, K. O. (1994). *Local knowledge of farmers in Ghana: Implications for extension and sustainable development programs*. Unpublished doctoral dissertation, Cornell University, Ithaca, New York.

Dei, G. J. S., B. L Hall, & D. Goldin-Rosenberg (2000). Introduction (pp. 3–17). In G. J. S. Dei, B. L. Hall, & D. Goldin-Rosenberg (eds.), *Indigenous knowledges in global contexts: Multiple readings of our world*. Toronto, Ontario: University of Toronto Press.

Deshler, D. (1996, April). External and local knowledge—Possibilities for integration. *Africa Notes*, 1–3.

Dewan, A. K. (1990). *Class and ethnicity in the hills of Bangladesh*. Unpublished doctoral dissertation, Department of Anthropology, McGill University, Montreal, Quebec, Canada.

——— (1993). The indigenous people of the Chittagong hill tracts: Restructuring of political systems. In M. K. Raha & I. A. Khan (eds), *Polity, political process and social control in South Asia: The tribal and rural perspectives*. New Delhi: Gyan Publishing House.

Gamage, D. T. (2001). A monastery system of higher education: Twenty-three centuries of Sri Lankan experience. *Asian Profile*, 29(1), 31–41.

Gain, P. (1998). *The last forests of Bangladesh*. Dhaka: Society for Environment and Human Development (SEHD).

——— (ed.) (2000). *The Chittagong hill tracts: Life and nature at risk*. Dhaka: Society for Environment and Human Development (SEHD).

Haque, E. (1990). Tensions in the Chittagong hill tracts (pp. 43–65). In S. Haq & E. Haque (eds.), *Disintegrative progress in action: The case of South Asia*. Dhaka: Bangladesh institute of Law and International Affairs.

Heyd, T. (1996). Reaction to Dr. Agrawal's article. *Indigenous Knowledge Monitor*, 4(1), 1–4.

Howes, M., & R. Chambers (1980). Indigenous technical knowledge: Analysis, implications and issues (pp. 239–340). In W. Brokensha & O. Werner (eds.), *Indigenous knowledge systems and development*. Lanham, MD: University Press of America.

Kapoor, D. (2007). Subaltern social movement learning and the decolonization of space in India. *International Education, 37*(1), 10–41.

Khan, A. M. (1999). *The Maghs: A Buddhist community in Bangladesh*. Dhaka: University Press Limited.

Korten, D. (1990). Getting to the 21st century: Voluntary action and the global agenda. West Hartford, Connecticut: Kumarian Press.

Kothari, B. (1996). *Towards a praxis of oppressed local knowledges: Participatory ethnobotanical research in indigenous communities of Ecuador*. Unpublished doctoral dissertation, Department of Education, Cornell University, Ithaca, New York.

Miles, A. (2004). Women's work, nature and colonial exploitation: Feminist struggle for alternatives to corporate globalization. *Canadian Journal of International Development Studies*, Special issue, *21*, 855–878.

Mohammud, A. (2005). *Plunder of forest resources unabated in Rangamati, Bangladesh*. Retrieved from, http://www.sos-arsenic.net.

Mohsin, A. (1997). *The politics of nationalism: The case of Chittagong hill tracts.* Bangladesh, Dhaka: University Press Limited.

Mohsin, A. (2003). *The Chittagong hill tracts, Bangladesh: On the difficult road to peace.* Colorado: Lynne Rienner Publishers Inc.

Mohsin, A., & I. Ahmed. (1996) Modernity, alienation and the environment: The experience of the hill people. *Journal of Asiatic Society of Bangladesh,* 41(2), 265–286.

Norberg-Hodge, H. (1991). *Ancient futures, learning from Ladakh.* San Francisco: Sierra Club Books.

O'Sullivan, E. (n.d.). *What kind of education should you experience at a university.* Toronto, Canada: Ontario Institute for Studies in Education, University of Toronto. Retrieved from, http://www.taiga.net/eecom2001/papers/osullivan.pdf

Phillips, L., & S. Ilcan. (2004). Capacity-building: The neoliberal governance of development. *Canadian Journal of Development Studies, 25*(3), 393–409.

Rafi, M. (2001). Immigration and political development history (pp. 21–32). In M. Rafi & A. M. R. Chowdhury (eds.), *Counting the hills: Assessing development in Chittagong hill tracts.* Dhaka: United Press.

Rahman, A. M. (1995). Participatory development: Toward liberation or co-option? (pp. 24–32). In G. Craig & M. Mayo (eds.), *Community empowerment: A reader in participation and development.* London: Zed Books.

Rahman, A. M., A. Khisa, S. B. Uddin, & C. C. Wilcock (2000). Indigenous knowledge of plant use in a hill tracts tribal community and its role in sustainable development (pp. 75–78). In P. Sillitoe (ed.), *Indigenous knowledge development in Bangladesh: Present and future.* Dhaka: University Press Limited.

Rashid, H. (2009, June 30). Tipaimukh Dam. *The Daily Star (Dhaka), 19*(165), 10.

Roy, A. (2008). *The shape of the beast.* Delhi, India: Penguin and Viking.

Samad, S. (1994, November 12). Environmental refugees in CHT. *Weekly Holiday,* 2.

———— (2000). *Dams caused environmental refugees of ethnic minorities: Hill Tracts of Bangladesh,* Serial NO SOC032. Retrieved from, http://www.dams.org/kbase.submissions/showsub.php?rec=SOC032

Semali, L. (1996). Reaction to Dr. Agrawal's article. *Indigenous Knowledge Monitor, 4*(1), 12–15.

Shiva, V. (1989). *Staying alive: Women, ecology and development.* London: Zed Books.

———— (2000a). Foreword: Cultural diversity and the politics of knowledge (pp. vii–x). In G. J. S. Dei, B. L. Hall, & D. Goldin-Rosenberg (ed oronto, Ontario, Canada: University of Toronto Press.

Shiva, V. (2000b). *Tomorrow's biodiversity.* New York: Thames and Hudson Inc.

Shoenu, M. (2007). Marma Utshob O Bibah [Festivals and marriage of the Marma community]. In S. Chakma (ed.), *Parbarto Chotrogram Upojati Uthshob O Bibah.* [Tribal festivals and marriage of the Chittagong Hill Tracts]. Rangamati, Bangladesh: Tribal Cultural Institute.

Talukder, S. (2006). *The Chakma race*. Rangamati, Bangladesh: Tribal Cultural Institute.

Tripura, P. (2000). Culture, identity and development. In P. Gain (ed.), *The Chittagong hill tracts: Life and nature at risk* (pp. 97–105). Dhaka: Society for Environment and Human Development.

Warren, D. (1991). *Using Indigenous knowledge in agricultural development*. Washington, DC: The World Bank.

United Nations (1990). *Report of the world summit for social development*. Copenhagen, Denmark: Author.

UNESCO (1995). *The cultural dimension of development: Towards a practical approach*. Paris: Author.

Zaman, M. Q. (1984). Tribal issues and national integration: The Chittagong hill tracts (pp. 311–324). In M. S. Qureshi (ed.), *Tribal cultures in Bangladesh*. Rajshahi, Bangladesh: Institute of Bangladesh Studies, Rajshahi University Press.

Zaman, M. Q. (1996). Development and displacement in Bangladesh: Toward a resettlement policy. *Asian Survey, 26*(7), 691–703.

ANIMALS, GHOSTS, AND ANCESTORS: TRADITIONAL KNOWLEDGE OF TRUKU HUNTERS ON FORMOSA

SCOTT SIMON

LIFE PROJECTS ON THE MARGINS OF FORMOSA

The Western Pacific island of Formosa, historically claimed by both Japan and China, is usually depicted as "Taiwan" in narratives about its economic accomplishments, high-tech industry, and robust democracy.[1] It is portrayed as a "laboratory of Chinese culture," yet also as a "laboratory of identities" and conflicting nationalisms (Corcuff, 2002, p. xxiii). After nearly four centuries of Chinese settlement, the dominant social and political groups on the island are strongly influenced by southern Chinese institutions and norms. Yet the craggy highlands and windswept coastlines of the island are inhabited by small communities of Austronesian peoples who have struggled to maintain their livelihoods amidst social dislocations caused by the imposition of Japanese and Chinese colonial regimes. They are Chinese only in the sense that Chinese citizenship was imposed on them by Chiang Kai-shek's Republic of China (ROC) after World War II (WWII).

Archaeological evidence suggests Austronesian inhabitation of Formosa for at least 6000 years. Formosa is recognized as the ancestral homeland of the Austronesian peoples, who range from Formosa in the north to New Zealand in the south, from Madagascar in the west to Easter Island in the east (Bellwood, Fox, & Tyron, 1995). The Austronesian peoples of Formosa in 2008 had a population of 494,107, or 2.1 percent of Taiwan's entire population. They are classified into fourteen officially recognized tribes: Atayal, Truku, Seediq, Saisiat, Amis, Tsou, Thao, Bunun, Rukai, Paiwan, Puyuma, Kavalan, Sakizaya, and Ta'u. The Truku, classified by the state until 2004 as part of the Atayal tribe, had a population in January 2009 of 24,600 people located largely in three townships on the East Coast and one in the Central Mountain Range (Indigenous Peoples Council [IPC], 2009).

From a development perspective, the indigenous people are marginalized. In 2007, they had an unemployment rate of 4.62 percent, compared to 3.83 percent in the general population. The general population earned 1.3 times the average monthly indigenous income, which was NT$17,156 (IPC, 2008), or US$500. Indigenous workers were 9.4 times more likely to work on temporary contracts than nonindigenous workers (IPC, 2008). Infant mortality rates (40.9 per 100) are more than double that of the general population (17.5 per 100), and the life expectancy of indigenous men (59.2 years) is 13.5 years shorter than that of nonindigenous men (Wen, Tsai, Shih, & Chung, 2004). Yet the indigenous peoples remain proud of their ways of life and are determined to keep intact their distinct lifestyles and languages. It is thus useful to distinguish between development and indigenous life projects:

> Life projects are embedded in local histories; they encompass visions of the world and the future that are distinct from those embodied by projects promoted by state and markets. Life projects diverge from development in their attention to the uniqueness of people's experiences of place and self and their rejection of visions that claim to be universal (Blaser, 2004, p. 26)

Higher unemployment rates and lower wages indicate racial discrimination and social exclusion. They also reflect choices not to migrate to cities in search of urban employment, as such strategies conflict with other possible life projects. Many villagers return from urban employment testifying that migration is not necessarily profitable due to higher costs of urban rent, food, and living expenses. They value the social networks of clan and community in the village, as well as the generosity of nature to provide food. The men can then invest their time and efforts in other life projects of hunting and community work. However, hunting is now illegal, which makes this once valued activity a conflicting source of pride and pain. This certainly contributes to the heavy drinking and lower life expectancy for indigenous men.

This chapter, based on eighteen months of field research between 2004 and 2007, examines hunting as a life project of Truku men and as a form of traditional knowledge about the mountain forests. How do Truku hunting practices embody traditional knowledge about law, human-animal relations, and the ancestors? How does hunting contribute to the continued learning of traditional values and language? How is this knowledge relevant to contemporary Taiwan and to the wider world?

INDIGENEITY IN TAIWAN

The legal term *indigenous peoples*, most recently codified in the United Nations Declaration of the Rights of Indigenous Peoples, emerged from efforts to expand human rights in the United Nations organizations created at the conclusion of World War II. Throughout the world, indigeneity emerged in the aftermath of colonialism and genocide: "Indigenism is an identity, like that which unifies survivors of the Holocaust, grounded in evidence, testimony, and collective memory" (Niezen, 2003, p. 15). Each indigenous community has its own history of colonialism, resistance, and agency.

Taiwan, under martial law from 1947 to 1987, saw the emergence of political indigeneity as part of the democratization of the 1980s. The indigenous social movement began with the establishment of the Alliance of Taiwanese Aborigines (ATA) in 1984 (Allio, 1998). The Presbyterian Church of Taiwan (PCT) played a founding role in this organization and in subsequent demands for indigenous rights (Rudolph, 2003). One of their first demands was to change their collective name from mountain comrades (*shanbao*, 山胞) to indigenous peoples (*yuanzhu minzu*, 原住民族).

The political, civic, social, and economic rights of indigenous people (*yuanzhumin*, 原住民) were incorporated into the additional articles of the ROC Constitution in 1994. The IPC was created in 1996. In 1997, the Constitution was again revised to recognize indigenous peoples (*yuanzhu minzu*, 原住民族), *with the final—s* in English, effectively promising them collective rights. Subsequent legal reforms included numerous laws concerning indigenous identity and rights, culminating with the 2005 Basic Law on Indigenous Peoples (Simon, 2007). Indigenous hunting rights were included in the Basic Law, but most hunting practices were not decriminalized.

HISTORY OF TRUKU MIGRATION

The Truku trace their origins to the Central Mountain Range. According to legends, the first humans sprang from the center of a giant rock in those mountains at the beginning of time. Those who trace their origins to the upper reaches of the Karali River first settled in the original village, or Tarowan, in today's Nantou County. In subsequent migrations, they split into groups of Tek-Daya (Tkedaya), Toroko (Truku), and Tauda (Teuda) (Institute of Ethnology, 1935). In the resultant three dialects, they all called themselves Seediq, Sejiq, or Sediq—meaning human being (Institute of Ethnology, 1935). These groups, in search of game and fertile land, followed the rivers eastward toward the Pacific Ocean.

Among the nine Truku bands that followed the Takili River were the founders of the Xoxos and Skadang bands (Institute of Ethnology, 1935).

Xoxos, which means the sound of snakes or many snakes in Truku, is 915 meters above sea level. Xoxos people trace their settlement back to the nineteenth century, when their ancestor Udao Botto arrived there and decided with his brothers to clear land for farming (Mowna, 1977). Skadang, which means molar in Truku, is 1,128 meters above sea level. Skadang people trace their ancestry back to Batto Umao, who established a settlement at Putsingan in around 1817. In search of better hunting grounds, his sons Yibang Batto and Qaxo Batto arrived in the current location (Mowna, 1977). Elders say they chose the name Skadang because they found a human molar while digging to build their first homes.

In 1875, the Qing Dynasty set up outposts on the East Coast of Formosa, including at Hsin-cheng, to prevent Japanese and other foreign encroachment on the island. Hsin-cheng was located some six hours by foot from Skadang and Xoxos, mostly through snake-infested forested mountains. The Truku established barter relations with these first Chinese settlers (Mowna, 1977). Truku elders say that their parents and grandparents traded meat for salt, matches, and knives. The place at the foot of the mountains was called Bsngan. This comes from the root *bais* (pair or couple). The elders said that people felt it necessary to descend the mountain together, as it could be dangerous to meet the Chinese alone.

The mountain indigenous peoples were subjected to state administration only after the 1895 transfer of Formosa to Japan. For the first twenty years, the Japanese encircled the indigenous groups with military boundaries fortified with guard posts, electrified barbed wire fences, and even mines. The Truku resisted Japanese rule until the Taroko Battle. In 1914, Governor General Sakuma Samata launched the Five Year Plan to Pacify the Savages. In May 1914, Japanese troops attacked the Truku using more than 6000 soldiers equipped with modern military equipment and reinforced with aerial bombardment of mustard gas (Morris, 2004; Walis & Yu, 2002). Some 2000 to 3000 Truku youth fought back with bows and arrows and simple hunting rifles for nearly three months before surrendering. Sixty-one Japanese soldiers and fifteen policemen were killed, and 125 soldiers and twenty-five police were injured (Walis & Yu, 2002; see also Fujii 1997). Nobody seems to know how many Truku were killed. Skadang people relish the story of how they lured the Japanese troops into the narrow gorge of the Skadang River, only to cause an avalanche of rocks to kill them. They also claim to have killed Governor General Sakuma Samata, whom the Japanese say returned injured to Japan and died in his hometown of Sendai.

The Japanese encouraged most of the Truku to leave their mountain homes and take up agriculture. Twenty-eight villages, including Skadang and Xoxos, refused to move. The Japanese occupied Skadang and Xoxos, set up local men as chiefs to control the tribe, and began construction on a police station, school, and road (Mowna, 1978). The Japanese, following American precedents of Indian policy, established "savage reserves" or *banjin shoyōchi*, 蕃人所要地 (Yan & Yang, 2004). After the 1930 Musha Rebellion, in which Seediq warriors killed 134 Japanese at a sports event in Central

Taiwan (Ching, 2000), the Japanese began the period of Imperialization (*Kō minka*, 皇民化) to transform all Taiwanese into loyal subjects of the emperor. The Japanese encouraged the indigenous people to take Japanese names, wear Japanese clothing, pray in Shinto shrines, and even to serve in the war effort. They permitted the Truku to continue hunting, which from the Truku perspective was a condition for loyalty. Many older Truku people still remember the Japanese period nostalgically, as they were permitted to hunt.

After WWII, sovereignty of Taiwan was transferred from Japan to the ROC and subsequently administered by the Chinese Nationalist Party (KMT). From 1952 to 1962, the government implemented a campaign to improve the livelihoods of mountain people. All indigenous Formosans were forced to learn Chinese, take Chinese names, and adopt Chinese lifestyles. They were again encouraged to abandon hunting and take up agriculture (Walis & Yu, 2002). In 1962, when the ROC was still a member of the UN, the government signed the 1957 International Labour Organization (ILO) Convention No. 107 Concerning the Protection and Integration of Indigenous and Other Tribal and Semi-Tribal Populations in Independent Countries (Iwan, 2005). In 1966, the government reorganized the reserves by registering land in the names of individuals rather than as land held collectively by bands. Indigenous individuals were permitted to sell or rent their property to other indigenous people, but not to nonindigenous people (Fujii, 2001). During this time, the Hsiulin Township Office assisted Skadang and Xoxos with the marketing of mushrooms, wild orchids, and even wild meat for a short period.

For Skadang and Xoxos, the most important land loss came with the establishment of the Taroko National Park on their territory. They were forced in 1980 to relocate to Minle District, Fushih Village, to make way for the park. With limited government subsidies, they constructed small homes in flood-prone land at the confluence of the river with the ocean. Although some individual landholders were compensated for their lands, they lost the rights to hunt, to cultivate crops, and to construct buildings even on land that remained registered in their names. Most of them still hold title to land within the park, but their economic activities are restricted by park regulations. The Taroko National Park opened in 1986. In 2004, Skadang and Xoxos established the Tongli Natural Ecological Autonomy Association to manage their relationship with the park administration. Throughout this long history, they have developed ways of relating with the environment, but also ways of advancing their own life projects in new political environments. They view political and economic change through their own moral perspectives.

OVER THE RAINBOW: SPIRITS AND THE LAW

Truku life involves a relationship between the living and the dead. Truku elders recount stories of the ancestors who founded their villages. Parents take their children up the mountains and put out stone markers identifying burial spots. Even before taking a sip of alcohol, Truku people pour some on the ground as an offering to the dead. The souls of the deceased become the ancestors (*utux*). The relations between the living and the dead, as well as between living people and their communities, and between people and animals, are all regulated by the moral code known as Gaya (Mowna, 1998; Walis & Yu, 2002).

These concepts maintain their original force in the wider community, especially when rituals are held to atone for violations of Gaya, but have also been reinterpreted by Truku Presbyterian pastors and elders. They have translated the name of the Christian deity as *Utux baraw* (the ancestral spirit above) and the Ten Commandments as the ten Gaya. Truku Presbyterians perceive Christianity as the extension of their own ethical system. Sometimes Truku people and even Taiwanese anthropologists (e.g., Yu, 1979) who work with them argue that Gaya has lost importance due to the impact of foreign political systems and the market economy. I argue that when Truku people make this argument, it constitutes resistance. They are taking the moral high ground and accusing the newcomers of destroying morality.

The Truku make frequent reference to the Rainbow Bridge or Spirit Bridge (*Hakaw utux*) and the need to follow Gaya in order to become an ancestral *utux*. As the story goes, men must hunt animals or human heads, and women must weave. After death, everyone must cross the Rainbow Bridge. At the mouth of the bridge is a giant crab that inspects the hands of the deceased. If a man has hunted or a woman has woven cloth, his or her hands will reveal permanent marks of blood as proof of hard labor. He or she is thus a real man or real woman (*sejiq balay*) and can cross the Rainbow Bridge to enter the land of the spirits. Those who do not reveal the bloody traces of labor will fall into the river below, where they risk being eaten by crabs. Those who cross to the other side become *utux* respected by their descendants and charged with enforcing Gaya.

In the past, Gaya included the enforcement of property rights. Some land, especially hunting territories (*kolonan*), was the collective land of the band (*alang*). Land under cultivation was family property. It was revered as the product of the labor of the ancestors, and could not be alienated for fear of spiritual retribution (Mowna, 1998). In the village, people still discuss this spiritual retribution, which they call *lumuba*. In 1980, when Skadang and Xoxos were forced to move, each landowner received NT$60,000 (US$1,764) and property in Minle. Some of the young people used the money to purchase motorcycles. However, after the trauma of the move, many of them had taken up alcohol and in a very short period of time, some twenty young people died in motorcycle accidents. People still attribute these deaths to the *lumuba* caused by giving in to state demands to leave the land. They emphasize that Truku warriors once protected their territory by hunting the heads of invaders. With the coming of the state and the market, some are willing to sell land to outsiders—at their own peril. Many of them are now landless people, whom traditionalists view negatively as *qnluli* (driftwood) or *ungat pusu* (rootless).

For many Truku men, the most painful aspect of political change is that they have lost the right to hunt. According to the 1972 National Park Law, hunting is forbidden within park boundaries. Those caught hunting must pay a fine of at least NT$30,000 (US$880), nearly two months of income, and risk imprisonment. In spite of this, hunting remains common. The difference is that a once valorized and prestigious activity must now be carried out in secret and is burdened with shame. Due to their hunting activities, Truku men have an intimate knowledge of wildlife. In the following section, I describe the most commonly hunted animals.

TRUKU HUNTERS AND THEIR GAME

Animals are central to Truku life and identity. Even small children who do not speak Truku know the names of forest animals. Truku people started teaching animal names to my two graduate assistants and I on the first day we arrived in the village. In good humor, they gave us names of animals as our nicknames, saying that we would have to prove through action our worth to acquire a human name. As the professor, I was given the name *Kumay* (bear), whereas my students were given the names *Pada* (a small muntjac deer) and *Rapit* (flying squirrel). They insisted that the face of one student was shaped like that of a muntjac, whereas the other had a fuzzy face like a squirrel. They said that a professor should be called Kumay, as bears are the strongest animals in the forest.[2] They also said that it is forbidden to hunt bears. If one kills a bear, one's entire family will also die due to spiritual retribution.

It is important to keep in mind that knowledge of nature is passed on through generations of hunting practices, and that this education happens in Truku language. Hunting is central to Truku identity. When the men return to the village with game, they cook the meat and serve it to others. Women are not permitted to touch hunting tools or knives used for cutting meat. The general name for wild animals and their meat is *samat*. Animals are hunted primarily by trapping and by rifle. In the past, bows and arrows or spears were used. All of these practices are referred to in Chinese as hunting (*dalie*, 打獵). In Truku, each form of hunting has its own name. Following and hunting an animal, for example, is called *rumuyuk*, whereas trapping is called *pha gasil* (Huang, 2000). This illustrates that hunting practices are much more developed among the Truku than among the agricultural Chinese. The weight of animals is important as they must be carried down treacherous mountain paths by hunters. A man who brings back a heavy animal gains prestige. When men hunt in groups, however, they never boast about who killed the animal. It is always described as a collective endeavor.

Since wildlife is so important to Truku identity and life projects, this section describes the most commonly hunted animals and their habits.

Wild boar (*bowyak*): This tusked boar is very common, and known for its fierce nature. Most hunters tell stories of being attacked by *bowyak*. Wild boars reproduce quickly, as their litters consist of up to twelve piglets. Like muntjac, they live everywhere in the forests from sea level to high mountains, with a preference for humid areas. They are attracted by human settlements, and hated for eating crops. They are hunted by rifle, dog teams, or trapping. Skilled hunters can identify their forest paths and determine where to lay traps. Wild boars can weigh up to 100 kilograms (Huang, 2000). Hunting *bowyak* requires great physical strength and contributes to the prestige of the hunters.

Mountain goat (*milit*): Mountain goats live in the forests at an elevation above 500 meters. They reproduce slowly, as they have a gestation period of 220 days and give birth to only one or two kids. They have a life expectancy of about fifteen years. Mountain goats are known for their ability to climb on rocky cliffs. It is said that *milit* identify hunters as their enemies and have been known to push hunters off cliffs to their deaths (Huang, 2000). They weigh about 30 kilograms. Hunters can identity their routes, and lay traps on narrow cliffs.

Muntjac (*pada*): These small deer live throughout the forests from sea level to the high mountains over 3000 meters. Their dog-like cries can often be heard in the forests at night. They weigh about eight to twelve kilograms. Hunters use traps, or chase them collectively with dogs (Huang, 2000).

Flying squirrel (*rapit*): Flying squirrels come in two subspecies: white-faced and red-faced. The white-faced squirrels live higher in the mountains at elevations of over 1000 meters, whereas the red-faced squirrels live in the foothills (Huang, 2000). Flying squirrels are usually hunted at night. Hunters shine headlamps into the forest so that they can identify squirrels from the reflections of their eyes. The most dangerous part of the hunt happens when the squirrel falls from trees into mountain crevasses and must be retrieved. Hunters search for them with the assistance of dogs. On a successful night, hunters can return with a backpack full of twenty to thirty kilograms of squirrels.

Macaques (*rowngay*): Macaques (a genus of Old World monkeys) live from 100 to 3000 meters above sea level. They love fruit, and often attack fruit trees planted for commercial purposes. They have a gestation period of 165 days, and give birth to one a year. They can live over thirty years (Huang, 2000). Like humans, macaques live in groups. The Truku hunt them with rifles or traps if they threaten their crops and fruit trees. Macaques can also be collateral damage, if they fall into traps meant to catch other animals. Many Truku say they prefer not to kill them, as they resemble human children.

Hunters are expected to share meat with others in the community, but still retain a great deal of power about how it should be distributed. Members of the immediate hunting group mix the fresh blood of larger animals with rice wine and drink it at the conclusion of a successful hunt. The raw liver (*rumul*) of muntjac, mountain goat, and deer are especially prized and shared widely with community members. As flying squirrels eat leaves, hunters only lightly boil the intestines (*iyaq*) and then eat them with the original content still inside. This is considered medicinal and is given to elders (*rudan*) as a sign of respect.

Hunters often cook and eat in the front of their homes, in which case they are expected to share with other community members or they sell the meat to Chinese middlemen. They also eat more privately with family and friends in the back yard or in the kitchen. They may also share meat strategically. One successful hunter/farmer, for example, always provided game and alcohol to community members and visiting cadres before meetings of the Farmers' Association. When he earned a good profit on his crop, other members of the community suspected it was because he had given meat to the cadres. Hunters say that the forest is their refrigerator, and emphasize the importance of sharing in Gaya. Those who eat alone are said to be selfish like rats (*qowlit*). This is considered morally reprehensible.

MORAL AND ECOLOGICAL DIMENSIONS OF HUNTING

Hunting is embedded in complex moral practices that involve relations with other community members, with the ancestral *utux*, and with the natural world. Successful hunters are said to have a certain spiritual character or power called *Bhring*. This may be limited to only one kind of hunting or one kind of animal. One who is skilled at hunting muntjac, for example, is said to have *Bhring* in relation to muntjac (*Bhring gmpada*). Hunters can increase their *Bhring* by carrying the parts of animals (a tusk or a paw) in a small hunter's sack (*lubuy*), which is never shown to others.

There are many taboos (*psaniq*) related to hunting. Hunters must avoid sexual contact with women before entering the mountains. They are not permitted to hunt at certain times in the life cycle of their families, such as when they are in mourning or when a woman is pregnant (Huang, 2000). Hunters enter the forest with an attitude of respect, avoiding laughing or singing. Sometimes they first make a ritual oblation of rice wine to the ancestors who once inhabited the forest. These taboos and practices illustrate how moral practices and hunting relate hunters intimately to the *utux*. The *utux* respond. The *utux* observe the traps and alert hunters through dreams when they should inspect their traps for game. If a hunter has violated moral principles, the *utux* will express their anger by causing a hunter to fail to catch prey, or even to get injured on the mountain. When hunters continually bring back prey and rest uninjured, this is evidence that they are morally upright men, real men (*snaw balay*).

There are also moral dimensions of the relationship between hunters and the environment. Before hunting, they may observe the behavior of the bird

Sisil (grey-eyed nun babbler). If this bird emerges from the right and cries, this is interpreted as a fortunate sign. If it emerges from the left and cries while flying in front of the hunter, this is considered unlucky and the hunter will go home (Huang, 2000). Many of these practices contributed to ecological sustainability in the past. In addition to the family circumstances that prohibit men from hunting, many hunters do not hunt in the spring when they may inadvertently kill a pregnant animal or the mother of a vulnerable young animal. Before the establishment of national parks, hunters hunted only in the territory of their own band (*alang*) and would not chase an animal into the territory of another *alang*. Some areas were known as sacred mountains (*pusu btunox*). As hunting was never done in these forested mountain areas, they served as conservation zones for the animals (Huang, 2000).

These practices allowed Truku men to hunt for millennia without depleting wildlife populations. It was only after the widespread destruction of habitat for reasons such as tea plantations, hydroelectric plants, forestry, and mining, that animals have become endangered and hunting identified as an ecological threat. Meat is now available through other means. Muntjac and wild boar are raised in captivity for their meat. Truku men, however, are reluctant to give up hunting. Hunting is a sign of masculinity, a source of prestige, and proof of one's moral standing. The sharing of game enhances one's social capital in the community. Some men, especially older and less educated men with little chances on the labor market, rely on hunting for their incomes. Their way of life has been criminalized, whereas even more destructive environmental practices are legal and encouraged by the state.

In an economic sense, the practice of hunting in today's society is an irrational choice, as individuals could earn more money at less risk of arrest and injury by raising some of the same animals in captivity or simply selling noodles in the night markets. Nonetheless, the *rudan* (elders) expect the younger men to provide them with traditional food and younger fathers enjoy teaching their sons to hunt. Hunting provides the most important environment where Truku children learn their native language and culture; the only other such environment being the Presbyterian Church. Even men who have migrated to the cities expect to hunt when they return home on weekends or on the holidays. All of this shows that hunting is a life project with important repercussions on personal pride and community survival.

The establishment of the Taroko National Park cast the final blow to sustainable Truku hunting practices. The previous practices by which members of different *alang* managed mutually exclusive hunting territories, reflecting the property rights regimes of Gaya, are no longer possible now that former hunting territories are managed by the State. Yet the State actively discourages hunting. This has led to a tragedy of the commons situation in which all hunters try to avoid the National Park police and no hunter seeks to exclude other hunters from any given territory. The taboo against chasing an animal into the territory of another *alang* is no longer relevant.

Even sacred mountains are now just places marked in Chinese on official park maps and known primarily as choice places to find rare animals beyond the usual park surveillance. Only the relative inaccessibility of these areas prevents them from being overhunted.

WRITTEN LAW: AN ALTERNATIVE TO GAYA?

To a certain extent, and after two decades of protest, lobbying, and negotiation, the State attempted to address this issue in the Basic Law on Indigenous Peoples. As a basic law it required all relevant laws and regulations to be redrafted and implemented, thus offering hope to indigenous people about the issue of hunting rights. Article 19 stipulated that:

> Indigenous persons may undertake the following non-profit seeking activities in indigenous peoples' regions: 1) Hunting wild animals; 2) Collecting wild plants and fungus; 3) Collecting minerals, rocks and soils; 4) Utilizing water resources. The above activities can only be conducted for traditional culture, ritual or self-consumption.[3]

In the absence of substantive legal revisions, some township governments took it upon themselves to begin taking registrations and giving permits for limited catches for cultural and ritual purposes. These were generally defined as festivals or rituals when local people might need two to three wild boars. They did not consider other cultural reasons such as courtship, nor did they give permission to individuals to hunt for subsistence purposes. Hunting is still prohibited in national parks, which means the entire hunting territory of Skadang and Xoxos. Trapping of any kind is still prohibited entirely. Therefore, police continue to arrest hunters. Even though the charges are consistently dropped in the courts, this angers hunters. Hunters express anger at the police, especially when they observe that policemen also hunt, or when the policemen arrest hunters and seize the meat for their personal consumption. In one story, a policeman confiscated a *milit* trapped by an elderly hunter, but was later seen eating it privately in his own home. When the brakes on the policeman's brand-new motorcycle froze and he died by catapulting off a cliff, this was interpreted as punishment by the spirits, his fatal *lumuba*.

In this context, hunting has become a rallying point for the indigenous social movements to express their concerns and for politicians of both major parties to mobilize indigenous support. This dynamic can be seen in both the KMT and the proindependence Democratic Progressive Party (DPP). In April 2007, for example, Skadang and Xoxos hunters protested at the Taroko National Park, which led to a public apology by the chief of police. KMT legislator Kung Wen-chi (Truku name: Yusi Dagun) subsequently held a public hearing at the Legislative Yuan encouraging revision of all relevant laws and suggesting that the DPP was not sincere about indigenous rights. Icyang Parod, DPP-appointed Minister of the IPC, said that it is the responsibility of the Legislative Yuan to pass relevant legislation, including that which will resolve the hunting problem. After the election of KMT Ma Ying-jeou as president in 2008, a series of indigenous protests against the new government used hunting rights as a rallying cry. The Basic Law clearly failed to meet the needs of indigenous hunters. This is because legislation is based on a political compromise between competing legislators rather than on indigenous values or the sacred law of Gaya.

CONCLUSION

In this chapter, I have used the Truku language to describe the Truku lifeworld of hunting, wildlife, and ancestors. I have done so in order to describe the non-Chinese cultural world in which they interpret their environment and rank different values. Truku understand their relation to the mountainous forests in terms of spiritually imbued relations with animals and ancestors. This is very different from the values of National Park administrators, the mining companies who are also able to negotiate access to parklands, or the legislators who have criminalized hunting. The values of those actors, who think of the forest and its resources in terms of tourist numbers, revenue, and profit, are not described in this chapter because they are well known to all readers of this book. Truku cultural values, which once enforced sustainable hunting and still emphasize sharing, stand in contrast to the values of the developers. Whenever Truku hunters emphasize Gaya, it is a form of resistance against this system that has marginalized them and their way of life. Hunting keeps Truku culture alive.

By no means do Truku hunters represent a disappearing way of life, even if the trope of the disappearing Indian has been part of common sense in both North America and Formosa for more than a century. It is certainly true that many small children in the villages speak Mandarin Chinese and only a few words of Truku. It is also true that some Truku men and women perform well in school, move to the cities, and enter into the mainstream of Taiwanese society. Nonetheless, for those who stay in the village, hunting and the Truku lifeworld embodied in hunting practices are very real aspects of daily living. Truku men still take their teenage sons into the mountains and impart to them knowledge of the natural world in the Truku language. I was able to learn Truku wildlife and hunting vocabulary only because these practices are still thriving. Those Truku individuals who perform well in school end up with degrees and urban lives, but sometimes return as aboriginal legislators, editors of Truku language materials, and advocates of indigenous rights. Those who do not perform well in school may get labelled as lazy or as alcoholics, but these are the ones who keep the language and hunting practices alive in daily life. The contributions and life projects of all these people, including the hunters, are needed in order to ensure Truku cultural survival.

A descriptive ethnography of Truku indigenous knowledge, learning, and life projects allows us to understand Asia in a new perspective. In the dominant narratives of East Asian development, Taiwan is one of the four Asian dragons that have prospered by embracing industrialization and free markets. Even this story has been eclipsed by that of a rising China that has embraced market-based reforms and is emerging as a global power in its own right. To many actors in academia, business, and diplomacy, even the claim of 23 million Taiwanese to sovereignty appears as an inconvenient annoyance. From this perspective, the existence of 24,600 Truku and probably less than 1,000 active Truku hunters must seem entirely insignificant. Outside of Taiwan, only an anthropologist could possibly be interested.

It is important to know, however, about this other set of Asian values. In a world where profit emerges as the sole criteria of value, Truku hunters are more concerned with sharing and community building. In a world of expanding state control over territories perceived as horizontal blocks, Truku hunters see their mountains vertically as animals and ancestors who inhabit different levels of the forest. In a world of written law that permits mining and criminalizes hunting in the very same mountains, Truku hunters assert the sovereignty of Gaya every time they lay a trap or shoot a squirrel. Truku hunters may be marginalized in Taiwan and in the world, but they have every right to be proud. Emphasizing differences between themselves and the world of states and markets, they dare to call themselves *Sejiq balay*, or real people. Their knowledge reveals the expansion of the state system as a historical process of colonialism, violence, resistance, and agency. It also gives us new knowledge about Taiwan and its place in the world.

Acknowledgment Note: The author gratefully acknowledges the Social Sciences and Humanities Research Council of Canada, as well as the generosity of the Truku and Seediq communities for making this research possible."

REFERENCES

Allio, F. (1998). L'autochtonie en terre taïwanaise [Indigeneity on Taiwanese Land]. *Recherches amérindiennes au Québec, 28*(1), 43–57.

Bellwood, P., J. Fox, & D. Tyron (eds.) (1995). *The Austronesians: Historical and comparative perspectives.* Canberra, Australia: Australian National University.

Blaser, M. (2004). Life projects: Indigenous peoples' agency and development. In M. Blaser, H. A. Feit, & G. McRae (eds.), *In the way of development: Indigenous peoples, life projects and globalization* (pp. 26–46). Ottawa, Canada: International Development Research Centre.

Ching, L. (2000). Savage construction and civility making: The Musha incident and colonial representations in colonial Taiwan. *Positions, 8*(3), 795–818.

Corcuff, S. (2002). Introduction: Taiwan a laboratory of identities. In S. Corcuff (ed.), *Memories of the Future: National Identity Issues and the Search for a New Taiwan* (pp. xi–xxiv). Armonk, NY: M. E. Sharpe.

Fujii, S. 藤井志津枝 (1997). 理蕃：日本治理台灣的計策 沒有砲火的戰爭(一) [Managing Savages: Japanese Policies for Administering Taiwan, a War without Artillery Fire, Vol. 1]. 臺北市：文英堂出版社 [Taipei: Wenyingtang].

Fujii, S. (2001). 臺灣原住民史：政策篇(三) [Taiwanese Aboriginal History: Policy, vol. 3]. 南投：臺灣文獻館 [Nantou: Taiwan Wenxian Guan].

Huang, C. H. 黃長興 (2000). 東賽德克群的狩獵文化 [The Hunting Culture of the Eastern Sejiq Group]. 民族學研究資料彙編 [Collection of Ethnological Research Materials] (15), 1–104.

Institute of Ethnology 土俗人類学研究室調查 (1935). 台湾高砂族系統所属の研究 [The Formosan native tribes: A genealogical and classificatory Study]. 台北：台北帝国大学 [Taihoku (Taipei): Taihoku Imperial University].

Indigenous Peoples Council (IPC) 行政院原住民族委員會 (2008). 96年原住民就業狀況調查與政策研究—就業狀況調查報告 [2007 Investigation and Policy Research on Aboriginal Employment Conditions: Report on the Employment Conditions Investigation] 臺北: 行政院原住民族委員會 [Taipei: Indigenous Peoples Council].

———. (2009). 原住民人口數統計資料 [Materials on Aboriginal Population Statistics]. Retrieved from, http://www.apc.gov.tw/main/docDetail/detail_TCA.jsp?isSearch=&docid=PA000000002833&cateID=A000297&linkSelf=161&linkRoot=4&linkParent=49&url=

Iwan, N. 黃鈴華 (2005). 台灣原住民族運動的國會路線 [The Parliamentary Line of Taiwan's Indigenous Movement]. 臺北 :國家展望文教基金會 [Taipei: Guojia Zhanwang Wenjiao Foundation].

Morris, A. (2004). Taiwan's history: An introduction. In D. K. Jordan, A. Morris, & M. L. Moskowitz (eds.), *The minor arts of daily life: Popular culture in Taiwan* (pp. 3–31). Honolulu: University of Hawai'i Press.

Mowna, M. 廖守臣 (1977). 泰雅族東賽德克群的部落遷徙與分查 (上) [Migration and Dispersal of the Atayal Eastern Sejiq Bands (part one)]. 中央研究院民族研究所集刊 [Journal of the Institute of Ethnology, Academia Sinica] 44, 61–206.

———. (1978). 泰雅族東賽德克群的部落遷徙與分查 (下) [Migration and Dispersal of the Atayal Eastern Sejiq Bands (part two). 中央研究院民族研究所集刊 [Journal of the Institute of Ethnology, Academia Sinica] 45, 81–212.

———. (1998). 泰雅族的社會組織 [Social Organization of Atayal]. 花蓮: 慈濟大學原住民健康研究中心 [Hualien: Tzu Chi University Research Center on Aboriginal Health].

Niezen, R. (2003). *The origins of indigenism: Human rights and the politics of identity.* Berkeley: University of California Press.

Rudolph, M. (2003). *Taiwans multi-ethnische Gesellschaft und die Bewegung der Ureinwohner: Assimilation oder kulturelle Revitalisierung?* [Taiwan's Multi-ethnic Society and the Aboriginal Movement: Assimilation or Cultural Revival?] Hamburg: LIT Verlag.

Simon, S. (2007). Paths to autonomy: Aboriginality and the nation in Taiwan. In C. Storm & M. Harrison (eds.), *The margins of becoming: Identity and culture in Taiwan* (pp. 221–240). Wiesbaden, Germany: Harrassowitz.

Walis, N., & G.-H. Yu, 瓦歷斯.諾幹 與 余光宏 (2002). 台灣原住民史: 泰雅族史篇 [The History of Formosan Aborigines: Atayal Tribe]. 南投: 國史館台灣文獻館 [Nantou: Taiwan Historica].

Wen, C. P., S. P. Tsai, Y. T. Shih, & W. S. I. Chung (2004). Bridging the gap in life expectancy of the aborigines of Taiwan. *International Journal of Epidemiology, 33*(2), 320–327.

Yan A. C., & K. C. Yang, 顏愛竟 and 楊國柱 (2004). 原住民族土地制度與經濟發展 [Indigenous Peoples Land System and Economic Development]. 板橋市：稻鄉 [Banqiao, Daoxiang Publishers].

Yu, G. H. 余光弘. (1979). 東賽德克人的兩性關係 [Gender Relations of the Eastern Sediq People]. *Zhongyang yanjiuyuan minzuxue yanjiu jikan* 中央研究院民族學研究所集刊 [Journal of the Institute of Ethnology, Academia Sinica] (48, Autumn), pp. 31–53.

CHAPTER 7

DEVELOPMENT ENTERPRISES AND ENCOUNTERS WITH THE DAYAK AND MOI COMMUNITIES IN INDONESIA

EHSANUL HAQUE

INTRODUCTION

Indigenous people have, over several generations, developed a deep respect for and wealth of knowledge about the natural environment as peoples who live with and in nature. Some regard indigenous people as "shapers of environmental history" (Smith & Wishnie, 2000, p. 514), as indigenous value systems are often based on an intimate relationship with nature and the on-going protection of nature and biodiversity. Indigenous people are variously referred to in state-discourses as tribes, first peoples/nations, aboriginals, ethnic groups, *adivasi,* and *janajat,* while occupational terms like hunter-gatherers, nomads, peasants, are also some times used interchangeably with "indigenous people" (United Nations Permanent Forum on Indigenous Issues [UNPFII], 2007). According to Cobo (1987):

> Indigenous communities, peoples and nations are those which, having a historical continuity with pre-invasion and pre-colonial societies that developed on their territories, consider themselves distinct from other sectors of the societies now prevailing in those territories. They form at present non-dominant sectors of society and are determined to preserve, develop and transmit to

future generations their ancestral territories, and their ethnic identity, as the basis of their continued existence as peoples, in accordance with their own cultural patterns, social institutions and legal systems. (para 362)

This chapter considers the situation of indigenous people in Indonesia, highlights the role and nature of indigenous knowledge and analyzes indigenous encounters with the drivers of development enterprises. The focus is on two indigenous communities in particular—the Dayak in Kalimantan and the Moi in West Papau. It is argued here that these indigenous groups are being dispossessed of their traditional lands and are facing challenges to their identity, culture, and economic livelihoods due to rampant logging, agro-industrial developments (oil palm production), and extensive mining operations carried out by both local and foreign companies. The chapter briefly discusses the responses/struggles of the Dayak and the Moi and addresses issues pertaining to indigenous people's rights and interests.

INDIGENOUS KNOWLEDGE, LAND, FORESTS, AND DEVELOPMENT

The identities and cultures of indigenous people are strongly connected to their traditional/ancestral land and the natural resources they depend on (Pritchard, 2009). In fact, land is not only fundamental to the survival of indigenous people; it is also seen to bear a profound spiritual and cultural significance. Moreover, indigenous people regard themselves as an integral part of nature rather than perceiving nature as separate to them and therefore subject to human domination (Deruyttere, 1997).

Indigenous people subscribe to concepts/practices of development that are based on traditional values, visions, needs, and priorities (UNPFII, 2007). Although indigenous knowledge is often seen as static and unchanging, in reality it is fluid and is a transforming agent (Barua & Wilson, 2005; Ellen et al., 2000). It is, therefore, adaptable and attuned to indigenous values. According to Dei et al. (2000), indigenous knowledge refers to "traditional norms and social values, as well as to mental constructs that guide, organize, and regulate the people's way of living and making sense of their world" (p. 6). Indigenous knowledge is thus characteristically holistic, integrative, and situated within wider cultural traditions (Ellen et al., 2000). There are three important elements in indigenous knowledge. It is (i) the outcome of a dynamic system (*innovative talent of each indigenous people*); (ii) an indispensable part of the physical and social environment of traditional communities; and (iii) a collective well-being (Viergever, 1999; emphasis in original).

Since the 1980s there has been a radical shift in the thinking toward the rights of indigenous people and their access to natural resources. It is now widely recognized that indigenous people possess their own effective science and resource use practices (Sillitoe, 1998). Indigenous groups are fully capable of applying their knowledge to their livelihood and they have developed distinctive methods of conservation and sustainable management practices (Colfer et al., 1997; Gegeo, 1998).

The right to occupy and use land collectively is intrinsic to the self-conception of indigenous people and, generally, this right is vested in the local community, the tribe, the indigenous nation, or group (Stavenhagen, 2006). While such rights are acknowledged and protected by legislation in some countries, dominant economic groups often succeed in turning shared possession into private property. Indigenous territories all over the world are of interest to a host of international corporations for their natural resources (e.g., minerals, oil, hardwood, etc.). Subsequently, indigenous people have become the latest victims of globalized development and as a result, their very existence is seriously threatened (Stavenhagen, 2006).

The right to land is a central issue for indigenous peoples who have faced a history of dispossession, colonization, and degradation. The land question is a major source of friction between governments and indigenous communities. Traditionally, indigenous people have had exclusive jurisdiction and say over their lands and territories. But in modern times their uncontested rights have been contravened in different ways in different countries. This is despite the 1989 International Labor Organization (ILO) Convention 169 that urges states to respect indigenous lands and territories, and declares the right of indigenous people to control their natural resources—the convention seeks to protect indigenous people from manipulation and the appropriation of their lands (Gilbert, 2006). That being the case, this convention has not yet been ratified by any country in Southeast Asia and the indigenous people in this region continue to lose their grip over land and resources because in most Southeast Asian states indigenous land rights are either generally unguarded or weak, and many indigenous people are threatened by logging, mining, land clearing, fire, or infrastructure programs (such as dams and roads) undertaken by national governments (Poffenberger, 2006; Stavenhagen, 2006; Xanthaki, 2003).

Indigenous people's strong bond with land is inspired by a sense of spiritual value. The special ties and attachment indigenous people have with land and nature justify notions of collective ownership of land. Furthermore, land is not a "commodity" that can be purchased or sold, but rather a substance endowed with sacrosanct meanings that define their own survival and identity. Similarly, the trees, plants, animals, and fish, which inhabit the land, are not "natural resources," but highly personal beings that constitute part of the social and spiritual universe (Davis, 1993, p. 1). Land, therefore, occupies a central place in the indigenous worldview—it is the land that owns people and constructs identities for them.

For forest-dependent people, forests ensure sustenance and subsistence. The intense symbiotic relationship with the forest, over millennia, has shaped indigenous societies, worldviews, knowledge, cultures, spirituality, values, and cultural identities. Their spiritual and customary laws, and sophisticated land tenure (mostly communal ownership) and resource management systems ensure that their needs are met and that the forests are adequately protected from demolition (Tauli-Corpuz & Tamang, 2007).

In many parts of the world forests remain the domain of the state and the rights of forest-dwelling people are either denied or remain unprotected. It is estimated that between 2000 and 2005 the annual net loss of forest area was 7.3 million hectares (Food and Agricultural Organization [FAO], 2005). The Intergovernmental Panel on Forests (the precursor of the United Nations Forum on Forests) reported that one of the main causes of deforestation and forest degradation was the failure of governments and other institutions to recognize and respect the land rights of indigenous people and other forest-dependent people (Tauli-Corpuz & Tamang, 2007).

Forest-based goods and services are central to the economic, social, and cultural rights of hundreds of millions of people around the world. The World Bank estimates that 90 percent of the 1.2 billion people living in extreme poverty depend on forest resources for some part of their livelihood (UNDP et al., 2005). In Indonesia, for example, more than ten million poor people live in state forest zones with good forest cover, while millions more depend on forest for their livelihood (Wollenberg et al., 2004). Indonesia has the second most ecologically diverse rainforests in the world occupying 10 percent of the world's rainforests and 40 percent of Asia's rainforests. The Indonesian archipelago contains 120.35 million hectares of forest, which is the largest forest area in Southeast Asia and the world's third largest after the Amazon and Congo Basins. In many places, these forests are home to indigenous cultures and people.

Although, various instruments of international law—including the United Nations Declaration on the Rights of Indigenous Peoples and the Convention on Elimination of All Forms of Discrimination—recognize the right of forest people to "own, control, use and peacefully enjoy their lands, territories and other resources, and be secure in their means of subsistence" (Colchester, 2007, p. 5), in many countries current forest governance regimes do not uphold international human rights standards. The main causes of deforestation and degradation include need and greed. Much of forest destruction is driven by commercial-scale economic activities that enjoy implicit and explicit state subsidies as it is in the case of Indonesia.

INDIGENOUS PEOPLE IN INDONESIA AND GOVERNMENT POLICY

Indonesia has an estimated population of 220 million and the government officially recognizes the existence of 365 ethnic and subethnic groups as *komunitas adat terpencil* (geographically isolated customary law communities) in the country. They number about 1.1 million. However, many other people consider themselves, or are considered by others, as indigenous people. The nation-wide indigenous people's voice, *Aliansi Masyarakat Adat Nusantara* (Alliance of Indonesian Indigenous People) (AMAN), uses the term *masyarakat adat* (traditional people) among other terms (see, for example, Perkumpulan Sawit Watch et al., 2007) to denote indigenous people. It characterizes indigenous people as groups of people having a historical continuity that developed in a given geographical area, and having their own values, ideologies, and economic, cultural, and social systems, as well as territories. The total number of indigenous people in Indonesia ranges between 30 million and 40 million (Nababan & Sombolinggi, 2009).

Indonesia's ethnic plurality is regarded as an asset of cultural riches supporting state unity, which is reflected in the oft-repeated national slogan, *Bhinneka Tunggal Ika* (Unity in Diversity). However, Indonesia has had problems recognizing the rights of indigenous people. During the New Order era (1965–1998), *tunggal ika* (unity)—which usually was understood as unified, standardized effort—was much strongly emphasized than *bhinneka* (diversity). Although in the postindependence period the Indonesian state recognized customary law and the self-governance of indigenous communities, it was rescinded with the introduction of the Local Administration Act of 1979, which replaced the wide variety of the people's own customary institutions with new, uniform administrative units. As a result, customary institutions were disempowered and they lost recognition (Colchester, 2007). Efforts were made to limit the manifestation of ethnic identity through policies and programs on development emphasizing uniformity and solidarity. Hence, overplaying ethnic identity was considered perilous to national unity and integrity (Safitri & Bosko, 2002).

Government officials argued that the notion of indigenous people was not acceptable as almost all Indonesians (with the exception of ethnic Chinese) are indigenous and thus entitled to similar rights. Hence, the government discounted calls for special treatment of groups identifying themselves as indigenous with the aim of integrating these people into the social and cultural mainstream of the country. This was to be achieved by defacing traditional ways of life or by forced resettlement. Indigenous people's rights to land and resources were not recognized and they were considered illegal occupants of state forest land which the government wanted to open up for logging, mineral extraction, plantation, and transmigration projects (the settlement of hundreds of thousands of migrants in indigenous people's lands in "outer islands"—Kalimantan, West Papua, and Sumatra).

After the fall of New Order government, in 1999 the AMAN posed a grave challenge to the Indonesian state: "we will not recognize the State, unless the State recognizes us" (World Agroforestry Center [ICRAF] et al., 2003, p. 1). However, the second amendment to the 1945 constitution of Indonesia in 2000 acknowledged indigenous people's rights. Provinces and districts can now pass laws recognizing customary institutions, but these are yet to be passed in more than a few areas. Moreover, the qualified recognition of indigenous people in the constitution, which makes guarded reference to the need to recognize them, "as long as they still exist" (Colchester et al., 2006, p. 47) weakens indigenous communities' capabilities to assert their rights.

The constitution gives the state "controlling power" to allocate land and natural resources in the national interest, while the Basic Agrarian Law of 1960 (Act 5) upholds customary law as long as it does not "contradict national and State interests, based on national unity and Indonesian social-ism" (Colchester et al., 2006, p. 49). The government interprets "national interest" as including all projects mentioned in national five-year plans and all areas zoned for development or conservation in Provincial Spatial Planning exercises. Indigenous people's rights in Indonesia are thus giving way to log-ging, oil palm plantations, dams, and mines.

Recently, Government Regulation No. 2 (2008) was passed regarding types of tariffs and state revenues from activities within forest areas that lie outside the jurisdiction of the Ministry of Forestry. The regulation allows protected forest zones to be utilized for mining operations. Paradoxically, many mining development sites are in forestlands and these rainforests play a vital role in protecting the country's watersheds, averting land degradation by soil conservation, reducing sedimentation, and ensuring a sustained flow of river water for drinking and irrigation.

Moreover, the New Indonesian Mineral and Coal Mining Law was adopted in 2008 to replace the old Mining Act of 1967, which had been used to facilitate the incursion of mining companies into indigenous ancestral lands, causing land conflicts and gross human rights violations. However, the new law fails to recognize the land rights of indigenous people and ignores their environmental concerns. Thus, it does not provide full protection of affected communities from the human rights abuses perpe-trated by mining giants to the country. The new law also does not leave any room for renegotiation of existing mining contracts, thereby promoting the continuation of an extraction model, similar to the previous Mining Act (Murdiyarso, 2008; Nababan & Sombolinggi, 2009). Indigenous lands, therefore, continue to face immense threats from the influx of extractive industries. Indonesia's abundant reserves of coal, copper, gold, tin and nickel has lured mining giants in the country. Such companies exercise monopoly power over their areas of operation and they have tremendous influence in the mining sites as they are often the only source of steady employment and infrastructure development.

THE DAYAK COMMUNITIES IN KALIMANTAN

Dayak is a generic term used to categorize a sizeable group of indigenous people of the island of Borneo. The island is divided between three countries: Indonesia, the Malaysian Federated States of Sabah and Sarawak, and Brunei. Indonesian Borneo or Kalimantan (Indonesian Dayak territory) has copious natural resources such as dense rainforests, large deposits of gold, coal, petroleum, and natural gas and it is divided into four provinces: West, Central, East, and South Kalimantan. Dayak, meaning "people of the upstream," are indigenous non-Malay people who live in the interiors of Kalimantan and in Sabah and Sarawak in Malaysian Borneo. There are between 2 million and 4 million Dayaks in Indonesia (Minority Rights Group International [MRGI], 2009). These indigenous inhabitants prefer to be identified as Dayak along with their specific tribal names such as Kayan, Kenyah, Benuaq, Merab, Lun Dayeh, and Punan (Joshi et al., 2004).

The Dayak see themselves as part of the natural world and respect and protect nature and the environment from which they derive their livelihood. They developed sophisticated systems of sylviculture (the cultivation of forest trees), and land use adapted to the tropical forests and rivers in their environment. They created unique crop varieties and complex oral literatures (Masiun, 2000). Throughout their history the Dayak have relied on forest resources and their environment has shaped their culture and livelihood.

The Dayak have skills in hunting, collecting honey, wax, scented woods, nuts, and bird's nests. Most Dayak practice swidden agriculture, traditionally involving long and complex rotations of crops and trees on various patches of land. Some groups like the Kenyah historically inhabited swampy sites where they grew taro and nonirrigated rice (Joshi et al., 2004). As the Dayak possess centuries of farming experience within tropical rainforests, their resource-use systems are amazingly sophisticated in terms of the information and analytic procedures they employ in making management decisions. They are also democratic and consensual in the processes they use to decide resource allocation and use regulations (Poffenberger & McGean, 1993). Besides, the Dayak communities have excellent knowledge of plant usage, function, and regeneration. Their knowledge and involvement are key to sustainable development of these natural resources (Joshi et al., 2004). Indeed, the Dayak indigenous systems and practices sufficiently support the maintenance of biodiversity.

The sustainability of the forest ecosystem determines the coexistence of humans with nature. However, the overexploitation of forest and other natural resources and establishment of agro-industrial developments in Kalimantan ignores this aspect of Dayak sentiment and culture. Logging and mining industries have brought an irreversible change in the traditional land use systems in the Dayak land. Extensive logging and mining activities in the region endanger the traditional practice of harmony and sustainability causing social strife and environmental degradation (Joshi et al., 2004).

The massive destruction of Kalimantan's rainforests continues largely unabated, threatening Dayak traditions and livelihoods. The remaining forest cover of Kalimantan in 2005 was only 50.4 percent, according to the World Wildlife Fund for Nature (WWF), down dramatically from 74 percent in 1985 (cited in MRGI, 2009). John Bamba (2000), a prominent Dayak leader, presented a long list of grievances that this destruction has caused the Dayak:

> For many decades, the government's policies on forest management, religion, education, political activities, and other issues that directly touch the lives of the people created a disempowerment process. For the Dayak in Kalimantan, it caused the destruction of their environment, loss of their land and forest, disempowerment of their local *adat* (customary) institutions, destruction of their culture, and the creation of uncertainty with regards to their future as indigenous peoples in Kalimantan. (p. 40)

The impacts of logging operations are catastrophic. Logging activities have caused soil erosion, flood, and pollution. According to an expert,

> apart from the tangible effects of logging (hunger, river siltation, etc.), there are less tangible effects—the sounds and smells of the forest, coolness and heat, sunlight, vegetation, and mud. The words and images, the Dayaks employ are contrastive and tinged with nostalgia: what the forest was like before and after logging.
>
> Brosius, 1997, p. 473

Moreover, poor logging practices mobilize debris that not only find their way into the streams and rivers but also into the marine environment where it damages mangroves and coral reefs, habitats crucial for aquatic life. The logged over areas are also very prone to forest fires (Bamba, 2000). In the 1980s and 1990s fire raged in the forests of East Kalimantan that caused catastrophe to the ecology and economy of the region, disrupting soil and water conservation and destroying germplasm resources; including plants and animals useful to mankind (Thompson & Duggie, 1996). Intensive logging activities in East Kalimantan have turned its forests into a fire hazard.

Of late, the government has invited investors to convert the West Kalimantan rainforests into oil palm and industrial tree plantations. Between 1984 and 2005 the area under oil palm plantations in West Kalimantan increased manifold, from 5,000 hectares to 382,000 hectares (Potter, 2008). Evidently, the main reason for the rapid expansion of oil palm plantations is that they provide rich bonanzas to domestic and international plantation owners and investors. To the Indonesian government, oil palm or biofuel is the green dollar tree (Wakker, 2006). In 2006 the government announced a plan to establish massive oil palm plantations in an area stretching 850 km in West Kalimantan along the Indonesia-Malaysia border. This area covers parts of the ancestral territory of 1 million to 1.4 million Dayak people. Under the Kalimantan Border Oil Palm Mega-Project some eighteen separate oil palm plantations have been proposed each with an average size of 100,000 hectares (Perkumpulan Sawit Watch et al., 2007). This mega project has been launched under the banner of "bringing prosperity, security and environmental protection to the Kalimantan border area"(World Rainforest Movement [WRM], 2006) and it is being implemented by PT Perkenunan Nusantara, the Indonesian State Plantation Corporation. It requires 18 million hectares of land. The advocates of oil palm plantations claim that this will reduce unemployment, alleviate poverty and bring environmental benefits. Yet this project comes with huge social and environmental costs—it will threaten Dayak identity by dislocating 300,000 Dayak people and will eventually reduce them to plantation laborers.

Industrial operations are the sources of some of the most toxic pollutants and the most seriously polluted environments in Indonesia. Mineral extraction is problematic. Mines can have adverse impacts on environment over long distances (through water or air pollution) and because they are often in remote locations, they can have hazardous effects on local people who remain uncompensated (Adams, 2001). Coal and gold mining in East Kalimantan and West Kalimantan causes severe environmental damage. Extraction activities create huge holes that are difficult to fill even after reclamation. The holes consist of high concentrates of by the mining companies acid that pollute the soil (Forqan, 2006). Large tracts of indigenous land have been expropriated by the mining companies, destroying tropical rainforests. Indigenous people and conservation NGOs have protested against the government's granting facilities to the resource extraction companies and the Provincial Legislative Assembly of East Kalimantan demanded the cancellation of all permits of coal mining in protected forests (Mines and Communities [MAC], 2004). Despite these efforts, both legal and illegal mining have continued to be rampant.

Conflicts and clashes between the Dayaks and logging/mining companies started in the 1970s. Since then various forms of resistance—road blockades (often human barricades), attacks on production facilities, and protest rallies—were used in West Kalimantan (Pramono et al., 2006). Such opposition against infringements of human and civil rights and pollution have been met with heavy-handed treatment by government forces. However, in 1981, a group of Dayak educators, led by A. R. Mecer, established a social work foundation, popularly called Pancur Kasih [*Yayasan Karya Sosial Pancur Kasih*] (Davidson, 2008). It emerged to prop up Dayakness, elevate self-esteem, uphold Dayak interests, and prevent cultural erosion. In 1991 Pancur Kasih set up Institut Dayakologi that is avidly committed to the preservation of the Dayak territories and evolution of Dayak institutions, language, and culture. Their activities are aimed at empowering the Dayak people to shape their own futures by building on traditional Dayak values and institutions (Masiun, 2000).

For the Dayak, soil is their body, the river is their blood, and the forest is the breath of their life. These three elements give them an identity as Dayak people, give shape to their culture and beliefs, and also provide them with livelihoods (Bamba, cited in Natalia, 2000). Therefore, the government's nonrecognition of the land rights of indigenous people pits the Dayak against the government and logging/mining companies who are looking to exploit resources. In the mid-1990s, as commercial enterprises approached to grab indigenous lands, Pancur Kasih, in a massive drive to establish their rights, spearheaded a participatory mapping movement of customary resources and lands. This campaign still enjoys widespread support and trains not only Dayak communities but also other indigenous groups about the mapping process that can be used to negotiate land claims and resource rights with the government and development enterprises. For Dayak, this countermapping is an important tool for resistance that challenges state claims over indigenous lands (Natalia, 2000; Pramono et al., 2006).

THE MOI PEOPLE OF WEST PAPUA

West Papua (formerly Irian Jaya)—Indonesia's easternmost province—is a tropical island with primeval rainforests. The Moi indigenous people who number approximately between 10,000 and 13,000 inhabit more than 2 million hectares in the Sorong region in northwest Papua sometimes referred to as the Bird's Head (Wing, 1994; International Foundation for Election Systems [IFES], 2003). The Moi are sometimes known as the Mosana, Mooi, or Mekwei people. The land and the forest for the Moi is *Tamsini*, which means center or mother of life. Within this "mother," different organisms and plants grow, including sago palm trees, the staple food of the Moi people. Insects, wildlife, and humans live within and from this forest.

The Moi people depend on the rainforest for their survival. For generations, they have lived in the forest, using it for food, medicine, raw materials and their identity. They also strive to preserve and secure their land and way of life from major logging and oil palm schemes. Similar to Kalimantan, logging is responsible for environmental disaster in West Papua. Many other indigenous people of West Papua, including the Moi, are threatened as vast tracts of land have been granted as concessions to timber companies, a practice which has detrimental social and physical impacts (World Wildlife Fund [WWF], 2000). Forests stabilize the landscape that indigenous people have a deep historical relationship with, but logging entirely alters the landscape. Brosius (1997) observes:

> There is a strong coherence between the physical landscape, history, genealogy, and the identities of individuals and communities. With logging, the cultural density of the landscape—all those sites with biographical, social, and historical significance—is obliterated. Thus, logging not only undermines the basis of indigenous peoples' subsistence but also destroys the very things that are iconic of their existence as a society. (p. 473)

The most pervasive effects of logging are the reduced water holding capacity of the land and increased soil erosion from rain victimizing the Moi people.

To exploit the country's resources the government has given the construction of roads a high priority. Road construction has allowed companies and investors easier access to remote areas to do their business. The roads are built to allow mining/logging companies' trucks, bulldozers, cars, and other heavy equipments to carry out exploitation activities. According to one source:

> Logging roads are carelessly constructed leading to substantial soil erosion and consequent silting of rivers and irregularity of river flow. Roads are routinely built over minor streams; the result is a roadside string of standing pools, which produce unusually high concentrations of mosquitoes and present the threat of malaria and other diseases. . . . The heavy machinery destroys trees used by local people for food sources and traditional medicines
>
> Australia West Papua Association [AWPA], 1995

In addition to widespread logging, giant gold, copper and coal mining companies continue to operate on traditional lands in West Papua. Mining operations bring pollution and diseases as they discharge tonnes of toxic wastes and they also destroy indigenous people's farming practices. For years the rivers and streams have been polluted with chemical wastes, detergents, and coal residues.

Lately, the Indonesian government announced a plan to massively expand oil palm plantations in West Papua threatening the rich, biodiverse forests of the province. After years of extensive logging in Sumatra and Kalimantan, West Papua is the only province in Indonesia that still has a vast amount of virgin forests. Greenpeace observes that there are 17.9 million hectares of forests intact in Papua and at least 9 million hectares or half of it is under serious threat from oil palm companies (cited in Singapore Institute of International Affairs [SIIA], 2007). This plantation crop has become a major source of tension and land conflict in West Papua.

The Moi people rose and united against big logging interests in their lands. As they built resistance, they have been labeled "security disturbers" (Wing, 1994, p. 12), an official term used to muffle any form of indigenous protest. The Moi have struggled for decades to gradually develop the grassroots base for a powerful social movement to assert their civil rights, and their rights to control their forests and waters. Like the Dayak, the Moi have come into conflict with the logging/oil palm industries operating there. They have deep grievances against these corporations and the government since such mega projects generate huge profits for a handful of people and the elites but are also not compensated for the Moi. Although the Moi are not as dominant as the Dayak are, they have demonstrated that they can organize on issues relating to their interests and make their voices heard.

One of the noteworthy conflicts has been between the Moi people and the PT Intimpura Timber Company. In 1990 the company was granted a logging concession of 339,000 hectares of ancestral Moi land by the government (Wing, 1994) without the consent of the traditional landowners. However, the Moi people fiercely resisted the encroachment of the company on their land, and made representations to the company, provincial government, forestry service and the army. But with the support of the local police force, the timber company kept logging the forest (Bahrumrum et al., 2006).

Under the "East Indonesia Development" drive the government opened up the province to outside exploration and in the process not only are the Moi people excluded from the decision to develop their lands but are also not compensated for the loss of land (Laine, 1992). Over the years, hundreds of Moi people defending and protesting against logging, plantations, and other harmful development activities on their customary lands and against infringement of their rights have been harassed, assaulted, intimidated, suppressed, and arrested by the authorities.

Now it is evident that the government and influential business interest groups have largely ruptured the Dayak and Moi people's traditional bonds to the lands and forests and often disregarded the spiritual and cultural ties people have with their lands and environment. It is evident from the above that logging activities, mineral developments, and oil palm plantations have become the focal points of environmental and social resistance movements and confrontations in the indigenous areas. Despite the economic benefits that may accrue for indigenous people, the impacts of such gigantic projects are ruinous and questionable. The worst scenario is that at times the commercial loggers, oil palm estates, and mining companies prefer to use migrant labor rather than employing local people, which results in social conflict and emotional damage for the people affected. In an asymmetric fight, the indigenous people often confront the collective strength of the development enterprises and the government and vow to continue their struggle despite the odds.

CONCLUDING REFLECTIONS

The Dayak and Moi people share a history of injustice. Their vision of prosperity reflects fundamental values that river, land, and forest are essential to their life and identity. To them, defending ancestral land, territory, and natural resources is therefore sacrosanct and inviolable. We know from the discussion that the areas covered by primary forests in Kalimantan and West Papua have been reduced drastically and that deforestation continues at an alarming pace. The government and development agencies have failed to gauge the environmental impacts of their policies and deeds and they have also been oblivious to the capacities and needs of those people who live in the areas they seek to develop. Clearly, there is an absence of a mechanism through which the government and business groups can consult/negotiate with indigenous people on development projects such as logging, mining, and oil palm plantations.

However, these are some of the challenges of harnessing Indonesia's vast forest resources to better contribute to growth, indigenous livelihoods, and environmental protection. Given the pressures on indigenous lands, resources, and ways of life, indigenous people have become strong advocates of environmental movements in the preservation of biological diversity and the sustainable management of fragile ecosystems. Indeed, the indigenous people of Indonesia are committed to protect and preserve the communal ownership of land, which, in their view, is inalienable and indivisible.

REFERENCES

Adams, W. M. (2001). *Green development: Environment and sustainability in the Third World.* London: Routledge.

Australia West Papua Association (AWPA) (1995). The environment: Resource boom or grand theft? Retrieved from, http://www.cs.utexas.edu/users/cline/papua/deforestation.htm

Bahrumrum, M., H. Prabowo, & H. Myrttinen (2006). *Struggling to survive: Kepa's impoverishment analysis—Indonesia.* Helsinki: Kepa.

Bamba, J. (2000). Land, rivers and forests: Dayak solidarity and ecological resilience. In J. B. Alcorn & A. G. Royo (eds.), *Indigenous social movements and ecological resilience: Lessons from the Dayak of Indonesia* (pp. 35–60). Washington, D.C.: Biodiversity Support Program, World Wildlife Fund (WWF).

Barua, B. P., & M. Wilson (2005). Agroforestry and development: Displacement of Buddhist values in Bangladesh. *Canadian Journal of Development Studies, 26*(2), 233–246.

Brosius, P. (1997). Prior transcripts, divergent path resistance and acquiescence to logging in Sarawak, East Malaysia. *Comparative Studies in Society and History, 39*(3), pp. 469–510.

Cobo, J. M. (1987). *Study of the problems of discrimination against Indigenous populations.* UN Doc. E/CN.4/Sub.2/1986/7. Geneva: United Nations.

Colchester, M., N. Jiwan, S. M. Andiko, A. Y. Firdaus, A. Surambo, & A. Pane (2006). *Promised land: Palm oil and land acquisition in Indonesia: Implications for local communities and indigenous peoples.* Bogor: Forest Peoples Program, Sawit Watch, HuMA and ICRAF.

Colchester, M. (2007). Beyond tenure: Rights-based approaches to peoples and forests: Some lessons from the forest peoples program. Paper presented to the *International Conference on Poverty Reduction in Forests: Tenure, Markets and Policy Reforms*, Bangkok, Thailand, September 3–7.

Colfer, C. J. P., N. Peluso, & C. S. Chung (1997). *Beyond slash and burn: Building on indigenous management of Borneo's tropical rain forests.* Bronx, NY: New York Botanical Garden Press.

Davidson, J. S. (2008). *From rebellion to riots: Collective violence on Indonesian Borneo.* Madison: University of Wisconsin Press.

Davis, S. H. (1993). Introduction. In S. H. Davis (ed.), *Indigenous views of land and the environment* (pp. 1–6). Washington, D.C.: The World Bank.

Deruyttere, A. (1997). *Indigenous peoples and sustainable development: The role of the inter-American development bank.* Washington, D.C.: IDB Forum of the Americas.

Ellen, R., P. Parkes, & A. Bicker (2000). *Indigenous environmental knowledge and its transformation: Critical anthropological perspectives.* New York: Routledge.

Food and Agricultural Organization (FAO) (2005). *Deforestation continues at an alarming rate.* Retrieved from, http://www.fao.org/newsroom/en/news/2005/1000127/index.html

Forqan, B. N. (2006). The dark picture of coal mining in South Kalimantan. Retrieved from, http://www.walhikalsel.org/eng/index2.php?option=com_content&do_pdf=1&id=55

Gegeo, D. (1998). Indigenous knowledge and empowerment: Rural development examined from within. *The Contemporary Pacific, 10*(2), pp. 289–315.

Gilbert, J. (2006). *Indigenous peoples' land rights under international law: From victims to actors.* New York: Transnational Publishers.

ICRAF, AMAN, & FPP (2003). *In search of recognition.* Bogor: World Agroforestry Center (ICRAF).

International Foundation for Election Systems (IFES) (2003). *Papua Public Opinion Survey-Indonesia.* Washington, D.C.: IFES.

Joshi, L., K. Wijaya, M. Sirait, & E. Mulyoutami (2004). *Indigenous systems and ecological knowledge among Dayak people in Kutai Barat, East Kalimantan: A preliminary report.* ICRAF Southeast Asia Working Paper. Bogor, Indonesia: World Agroforestry Center.

Laine, A. (1992). Moi resistance standing up to Indonesia. Retrieved from, http://multinationalmonitor.org/hyper/issues/1992/10/mm1092_07.html

Mines and Communities (MAC) (2004). East Kalimantan's parliaments disapprove mining in protected forest. Retrieved from, http://www.minesandcommunities.org/article.php?a=1835

Minority Rights Group International (MRGI) (2009). Dayak. Retrieved from, http://www.minorityrights.org/4435/indonesia/dayak.html

Masiun, S. (2000). Dayak NGO responses to national legal and policy frameworks affecting *adat* governance in Indonesia. Paper presented at the International Association for the Study of Common Property (IASCP) Conference. Bloomington, Indiana, May 31–June 4.

Murdiyarso, D. (2008, March 18). Indonesian forest law misjudges the climate. *Jakarta Post.* Retrieved from, http://www.minesandcommunities.org/article.php?a=8535

Nababan, A., & R. Sombolinggi (2009). Indonesia. In K. Wessendorf (ed.), *The indigenous world.* Copenhagen: International Work Group for Indigenous Affairs (IWGIA).

Natalia, I. (2000). Protecting and regaining Dayak lands through community mapping. In J. B. Alcorn & A. G. Royo (eds.), (pp. 61–72).

Perkumpulan Sawit Watch et al. (2007). Request for consideration of the situation of indigenous peoples in Kalimantan, Indonesia, under the United

Nations committee on the elimination of racial discrimination's urgent action and early warning procedures. *Committee on the Elimination of Racial Discrimination*, Seventy-First Session, 30 July–8 August.

Poffenberger, M. (2006). People in the forest: Community forestry experiences from Southeast Asia. *International Journal of Environment and Sustainable Development*, 5(1), 57–69.

Poffenberger, M., & B. McGean (eds.) (1993). *Communities and forest management in East Kalimantan: Pathway to environmental stability*. Berkeley, CA: University of California, Center for Southeast Asia Studies.

Potter, L. (2008). Dayak resistance to oil palm plantations in West Kalimantan, Indonesia. Paper presented to the *17th Biennial Conference of the Asian Studies Association of Australia*, Melbourne, July 1–3.

Pramono, A. H., I. Natalia, & Y. Janting (2006). Ten years after: Counter-mapping and the Dayak lands in West Kalimantan, Indonesia. Paper presented at the *11th Biennial Conference of the International Association for the Study of Common Property (IASCP)*, Bali, June 21–23.

Pritchard, R. (2009). Participation of indigenous peoples in natural resource development in Australia. Retrieved from, http://www.resourceslaw.net/documents/RIIndigenousParticipationMay09.pdf

Safitri, M. A., & R. E. Bosko (2002). *Indigenous peoples/ethnic minorities and poverty reduction in Indonesia*. Manila: Regional and Sustainable Development Department, Asian Development Bank.

Sillitoe, P. (1998). The development of indigenous knowledge: A new applied anthropology. *Current Anthropology*, 39(2), 223–252.

Singapore Institute of International Affairs (SIIA) (2007). Indonesia seeks international cooperation on reforestation and bird flu in Bali. Retrieved from, http://www.siiaonline.org/?q=programmes/insights/indonesia-seek-international-cooperation-reforestation-and-bird-flu-bali

Smith, E. A., & M. Wishnie (2000, October). Conservation and subsistence in small-scale societies. *Annual Review of Anthropology, 29*, 493–524.

Stavenhagen, R. (2006). Indigenous peoples: Land, territory, autonomy, and self-determination. In P. Rosset, R. Patel, & M. Courville (eds.), *Promised land: Competing visions of agrarian reform* (pp. 208–217). Oakland, CA: Food First Books.

Tauli-Corpuz, V., & P. Tamang (2007). *Oil palm and other commercial tree plantations, monocropping: Impacts on indigenous peoples' land tenure and resource management systems and livelihoods*. Working Paper. New York: United Nations Permanent Forum on Indigenous Issues (UNPFII).

Thompson, H., & J. Duggie (1996). Political economy of the forestry industry in Indonesia. *Journal of Contemporary Asia, 26*(3), pp. 352–365.

United Nations Development Program (UNDP), United Nations Environment Program, World Bank, World Resources Institute (2005). The wealth of the poor—Managing ecosystems to fight poverty. Washington, D.C.: World Resources Institute.

United Nations Permanent Forum on Indigenous Issues (UNPFII) (2007). Who are indigenous peoples? Retrieved from, http://www.wipce2008.com/enews/pdf/wipce_fact_sheet_21-10-07.pdf

Viergever, M. (1999). Indigenous knowledge: An interpretation of views from indigenous peoples. In L. M. Semali & J. L. Kincheloe (eds.), *What is indigenous knowledge: Voices from the academy* (pp. 333–343). New York: Falmer Press.

Wakker, E. (2006). *The Kalimantan border oil palm mega-project*. Amsterdam: AIDEnvironment.

Wing, J. R. (1994). The impact of development on the population and environment of the Indonesian province of Irian Jaya. Retrieved from, http://www.papuaweb.org/dlib/s123/wing/_ma.pdf

Wollenberg, E., B. Belcher, D. Sheil, S. Dewi, & M. Moeliono (2004, December). *Why are forest areas relevant to reducing poverty in Indonesia?* Governance Brief, No. 4. Bogor: Center for International Forestry Research (CIFOR).

World Rainforest Movement (WRM) (2006, June). Indonesia: A call to cancel plans to develop 3 million hectares of oil palm plantations. *WRM Bulletin*, Issue 107. Retrieved from, http://www.wrm.org.uy/bulletin/107/viewpoint.html

World Wide Fund for Nature (WWF) (2000). *Critical threats to Irian Jaya environment*. Retrieved from, http://members.tripod.com/wwfsahul_cs/threats.htm

Xanthaki, A. (2003). Land rights of indigenous peoples in South-East Asia. *Melbourne Journal of International Law 4*(2), pp. 467–496.

PART II

FORMAL EDUCATION

Rethinking and Reconstituting Indigenous Knowledge and Voices in the Academy in Zimbabwe: A Decolonization Process

Edward Shizha

Introduction

This chapter focuses on indigenous knowledge recovery in the academy as an anticolonial and antiracist project that needs to be pursued to liberate the minds of indigenous academics and researchers who have wittingly or unwittingly adapted Western knowledge as the universal knowledge and the panacea for underdevelopment in Zimbabwe. The anticolonial project evolves from centuries of colonialism's efforts to methodically eradicate indigenous ways of seeing and interacting with the world (Simpson, 2004). The colonial project was meant to subjugate and suppress indigenous people's historical commemorations (Shizha, 2008). In this project the academy was created as the epicenter of colonial hegemony, indoctrination, and mental colonization. The anticolonial project focuses on decolonization in academic settings in Zimbabwe, rupturing and challenging the current colonial politics of knowledge production and dissemination. The process includes, but is not limited to reclaiming, rethinking, reconstituting, rewriting, and validating the indigenous knowledges and languages, and repositioning them as integral parts of the academy in universities where teaching and learning reinforce hegemonic and oppressive paradigms which allocate differential social locations to Western and indigenous knowledges. The majority of indigenous academics/scholars in Zimbabwe are conservative and resistant to change and fear challenging the so-called dominant and/or "stable" knowledge. Indigenous academics or scholars in Zimbabwe are the "native" Africans, who also happen to be indigenous Zimbabweans, those who have their origins and cultural experiences based on their indigenous communities. However, indigenous academics or scholars, unlike the majority of indigenous Zimbabweans who live in rural areas, have had their learning or formal education shaped mainly by colonial experiences and colonial academic curricula. This chapter provides an argument for including or integrating indigenous knowledge and voices as constructions and constituents of academic knowledge and deconstructions of colonial educational discourse in higher education in Zimbabwe.

WHY STUDY INDIGENOUS EDUCATION IN <AN>ACADEMY IN ZIMBABWE

For almost a century, Zimbabwe was under the British colonial administration that defined formal knowledge. Education was narrowly defined in terms of assimilation of Eurocentric middle-class habitus. The hegemonic description of European knowledge as positivist and empirical, while indigenous knowledge and voices were demeaned as unscientific and irrational for social development, was designed to silence and "kill" the voices of the indigenous people in Zimbabwe (Shizha, 2007). In the process, indigenous historical commemorations were marginalized and colonized. What we know today is that social development cannot succeed without the agency of the people intended to benefit from it. Therefore, education, which is the vehicle for development, has to be democratized by reconstituting and transforming knowledge discourses and constructs in Zimbabwe. This can be done by undertaking inclusive projects and programs that focus on disrupting and deconstructing the Eurocentric hegemonic perspective that is central to academic knowledge constructs. The discursive project of "Indigenous knowledges" is seen as a way to rupture the sense of comfort and complacency in conventional approaches to knowledge production, interrogation, validation, and dissemination (Dei, 2002) in the Zimbabwean educational settings. Indigenous knowledge and voices should be integrated in the academic corridors in a manner that does not perpetuate their domination and recolonization. Through an antiracist and postcolonial approach, indigenous communities can contribute to the much needed voices in curriculum planning and knowledge dissemination. Generally, there is lack of initiative and agency to rethink and reshape the processes of delivering education in indigenous Zimbabwean contexts. Academic work is political, and Zimbabwean academics should see themselves as "agents of justice." Therefore, academics in Zimbabwe should decolonize or liberate their Western assimilated perspectives on knowledge production and dissemination to achieve social change and justice.

The problems that Zimbabwe faces today in restructuring its academic knowledge are partly embedded in the colonial legacy. For nearly a century, when Zimbabwe was under colonial rule, the majority of indigenous people had no say in or influence on government policies and political decisions that affected the education system (Zvobgo, 1996). Since indigenous people were oppressed and not politically empowered to make fundamental decisions affecting their education, racism and imperialism became the main cause of the indigenous Zimbabwean's problems. Formal academic education was a creation of a foreign "dominant" culture (Shizha, 2006a), formulated and structured around the nineteenth century British middle-class education system, which had a hegemonic and demonizing effect on the indigenous master narratives. The imposed hegemonic culture disrupted "the values of pre-settler and pre-colonial notions of learning . . . [that] were essential in reflecting the social and cultural needs and expectations of the community" (Abdi, 2005, p. 29). The arrival of European colonialism in Zimbabwe, as elsewhere in Africa, led to the perforce imposition of European or colonial worldview, which was largely responsible for not only the deliberate distortion of the traditional projects of education already in place (Rodney, 1982), but also of the indigenously based and comprehensive programs of development that were achieved and put in place over hundreds of years (Nyerere, 1968).

Without doubt, colonial education was a larger component of the colonial project to dehumanize Africans by imposing both inner and outer colonization (Shizha, 2005). Both inner and outer colonization were based on the premise that Africans would assimilate into the European life styles and values that were themselves a threat to the identity and self-perceptions of the indigenous people. To a greater extent, colonial education led to psychocultural alienation and cultural domination (Mazrui, 1993). Based on cultural imperialisms, indigenous Zimbabweans were defined and portrayed as inferior to Europeans and were deliberately taught to hate their cultural identities and to internalize the racial stereotypes of the colonizer. Moore (1997) argues that indigenous knowledges and identities do not reside in a fixed, static metaphoric site or space removed from practice, performance, power and process. By acculturating indigenous people, the settlers believed they were annihilating a static and fixed predisposition (Shizha, 2006b). In fact, because indigenous knowledges and identities reside within the "situated [political] practices through which identities and places are contested, produced and reworked in particular localities" (Moore 1997, p. 87), they were never obliterated and continue to exist to date. Nonetheless, European hegemony was and still has been about the ways in which cultures are constituted, about dominant and marginalized cultural narratives, selves, and identities. As Africans, we need to invent ways of rewriting or changing those dominant narratives and deconstruct "white" superiority and power.

The idea of assimilation is important when dealing with colonial education; and assimilation involves those who are colonized being forced to conform to the cultures and traditions of the colonizers. In colonial Zimbabwe, as elsewhere in Africa, colonial philosophy discounted African indigenous knowledge systems and cultural practices as invalid and irrelevant to colonial economic interests, and that through social and cultural assimilation, indigenous people would receive civility and enlightenment. Cultural assimilation is the most effective form of political action that was used by colonialists. To that extent, the greatest challenges that the indigenous Zimbabweans faced due to colonization and the "civilization" project were the violations of their human rights, knowledges of survival, rights to land, cultures and traditions, and the maintenance of a connection to the spiritual as well as to contemporary material realms of life (Dei, 2002). Because indigenous peoples' knowledges and culture were perceived as "barbaric" and "backward" the settlers overlooked and ignored their use and vitality in agricultural, pastoral, and other "conventional" land uses, and the value in environmental and biodiversity conservation, management, and sustainability (Shizha, 2006b).

The colonial educational project has not ended since it started despite the decolonization process that has taken place in most formerly colonized states. It has been repackaged in the form of the heinous and exploitative process of globalization. Globalization has meant profound changes to people's lives, some of which have been contradictory. Some of its influence has impacted on societal values and worldviews. The influence of globalization throws insurmountable and unprecedented challenges on inclusive academic planning and delivery that incorporates indigenous perspectives. Knowledge in higher academic institutions is dominated by so-called universal global knowledge, which is predetermined Western ideological persuasion. In this age of knowledge economy, universities and other institutions of higher learning play an important function in the development of any nation. They accomplish this function through the production of highly skilled manpower and the creation of new knowledge. However, if this knowledge is not culturally and socially specific, it might fail to fulfill the goals of national and social development.

Indigenous knowledge in the academy is faced by challenges as globalization tends to promote the pseudoscientific and modern way of knowing that displaces the Eurocentic worldviews. However, the interest that the international community is showing in indigenous ecosystems can stimulate the desire to globalize traditional ecological knowledge. Unfortunately, indigenous scientists are not taking enough advantage of this new interest in indigenous ecological knowledge, while concerns are raised about global biodiversity that exploits indigenous ecosystems. Western scientists, plant breeders, and biotechnologists are at the forefront of vandalizing and appropriating genetic material from indigenous communities (Shizha, 2008).

The appropriated information and indigenous materials are processed, pack-aged, and commoditized as Western products enforced by legal mandate and they are patentable excluding the indigenous owners from being identified with the final product. According to Mead (1994), the misappropriation of indigenous knowledge is escalating and is particularly virulent in key areas of research such as the environmental sciences and medicine. For example, the Washington-based group Future Harvest says that a US$220 million annual market for *Prunus africana* as a prostrate remedy could lead to extinction of the slow-maturing evergreen tree in the African wilds (Cappuccio, 2004). The externalization and appropriation of *Hoodia gordinii* and its international marketing as an appetite-suppressant is a disservice to the San people of Botswana, South Africa, and Namibia who have been disowned of their right to their material knowledge through patents as well as intellectual property rights that have excluded them. Indigenous academics in Zimbabwe should face blame for this situation. Very few of them make use of the indigenous ecological system to advance knowledge and material benefits from the tradi-tional ecological knowledge that surrounds them, thus leaving global rules on patenting of genetic resources via the World Trade Organization to privatize indigenous genomes, and the biodiversity upon which indigenous people depend (Shizha, 2008).

THE ASSIMILATED AFRICAN ACADEMIC

The encounter between the colonizer and colonized subjects disrupted ways of knowing, learning, and teaching for most indigenous peoples in the world. It also resulted in the erosion of cultures and ideas, and most importantly, the colonization of minds (Wane, 2008). The imputing of a Western psy-chological self was at the centre of academic colonization (Wilson, 2004). The colonial education system created an African scholar and elite who was divorced from his/her indigenous roots and one that lacked a holistic self-identity. Such scholars were assimilated into European culture and alien values and beliefs that created fragmented beings detached from their cul-tural communities. Cultural assimilation, the intended homogenization of cultural traits within selected individuals, contributed to the accumulation of colonial society-specific human capital, which was culturally alienated. Cultural alienation occurred when schooling replaced traditional indigenous cultures with Eurocentric ways of knowing and learning (Shizha, 2007). The process of academic colonization required the complete subjugation of minds and spirits so that political colonization and economic appropriation could be advanced with some indigenous people acting as willing accom-plices. Mental or psychological colonization was conducted through, among other mechanisms, Western education, texts, and literature. Some of the Zimbabweans who became "victims" of Western education became willing accomplices in perpetuating the cultural bomb enshrined in the colonial project. Ngugi wa Thiong'o (1986) describes the aim of the imperial cultural bomb as to "annihilate a people's belief in their names, in their languages, in their environment, in their heritage of struggle, in their unity, in their capacities and ultimately in themselves" (p. 3). The assimilated indigenous Africans became the colonial tools for reinforcing the commodities of colonial domination and exploitation. In the process, they poorly perceived themselves as purveyors of Western "civilization" and "enlightenment" and posed as the fiduciary of all knowledge (Shizha, 2008). Within the assimilationist perspective, the holistic and totality of social construction of knowledge was ignored. Assimilation disrupted the long-run benefits of greater cultural fluidity and flexibility.

Sociocultural theory assumes that learning cannot be dissociated from interpersonal interactions located in cultural frameworks (Bourdieu & Passeron, 1977). Socially situated learning recognizes that values, emotions, experiences, and cultural contexts are integrally related to learning (Chinn, 2007). Western approaches to knowledge decontextualize knowledge production and dissemination. In Zimbabwe, the newly created academics or the elite disregarded the recognition that cultural diversity is associated with diverse ways of understanding how people relate to each other and how knowledge is produced in diverse social and cultural contexts. If the new academics could realized that cultural diversity and diffusion produces alternative modes of knowledge production and understanding in society, they could generate greater fluidity and flexibility that enhances the accumulation of knowledge that is more widely applicable in Zimbabwe.

DECOLONIZING THE ACADEMY

Decolonization is "a social and political process aimed at undoing the multifaceted impacts of the colonial project and re-establishing strong contemporary indigenous nations and institutions based on traditional values, philosophies and knowledge" (Smith, 1999, p. 19). It is considered to be a broad concept of which anticolonialism is a part. The recovery of indigenous knowledge is deeply intertwined with the process of decolonization because for many indigenous scholars it is only through a consciously critical assessment of how the historical process of colonization has systematically devalued indigenous ways that they can begin to reverse the damage wrought from those assaults (Simpson, 2004; Wilson, 2004). The revaluing of traditional knowledge has to begin in indigenous communities and among indigenous people, not only because they are the major holders of the knowledge and the major impetus for academic and mental decolonization, but also so that they can prevent that knowledge from being appropriated by international vultures seeking to expropriate valuable indigenous ecological systems.

Academics in Zimbabwe need to be proactive and show interest in indigenous research that can facilitate indigenous peoples' struggle against the ravages of colonialism. Academic decolonization and assertiveness on the part of local scholars and researchers can facilitate the fight against further neocolonial encroachments (Semali & Kincheloe, 1999). As Adjei and Dei (2008) argue,

> what motivates us to write about decolonization in this historic moment is more than our experience. It is more of a political decision. Africa's issues are a matter of social justice and fairness. Africa has contributed a lot to the scholarship and material growth of the world, especially, in Europe and North America, yet Africa is often talked about as if the continent does not matter to the world. Even more troubling is the manner in which Africa has become a basket case for the intellectuals of the West. (p. 141)

Indigenous African scholars can emancipate themselves fully by taking their place in the global knowledge production and dissemination from their own African perspective. They should distance themselves from dominant theories that devalue their indigenous contributions to global knowledge. Decolonization plays a crucial role in positioning indigenous knowledges in the academy, thus placing indigenous knowledges into the arena of social development, while also deconstructing the false messianic complex that the West is the savior of African problems.

DECOLONIZING THE SELF

In order for educators to discuss ways of knowing, teaching, and learning for indigenous people, it is imperative that we ground our analysis in the history of those people, that history which connects the present with colonial and neocolonial pasts (Chabal, 1996). As indigenous educators we must realize that what is termed academic knowledge in Zimbabwe is rooted in a "white" cultural context and perceived as universally scientific. We need to interrogate our locations and perspectives in conventional academic knowledge. As reformist scholars, we should engage in an anticolonial resistance and transformative discourse as an educational experience. For example, anticolonial discourses within the context of attempting to reclaim indigenous African identities are best captured in the works of African writers, such as Fanon's (1967) *Black Skin, White Masks,* and Ngugi wa Thiong'o's (1986) *Decolonizing the Mind.* Scholars and academics in Zimbabwe should tenaciously adapt critical lenses and confront the negative representation of Africa and its people. Indigenous Africans suffered centuries of colonization and psychological repression that undermined their selves and their humanity. Ama Ata Aidoo (2000) captures the experiences of most indigenous Africans during and after colonialism:

> I grew up knowing that Europeans had dubbed Africa "The Dark Continent".... Since then its ugly odour has clung to Africa, all things African, Africans and people of African descent everywhere, and has not faded yet . . . I am not a psychologist or a psychoanalyst. However, I do know that it has not been

easy living with that burden. Africans have been the subject of consistent and bewildering pseudo scholarship, always aimed at proving that they are not inferior human beings. Even when there was genuine knowledge it was handled perniciously: by anthropologists and social engineers, cranial and brain-size scientists, sundry bell-curvers, doomsday, medical and other experts. (p. 1)

Wane (2008) observes that decolonizing the self is the most difficult process and that the indigenous elite has become a commodity of Western ideology. Most of us indigenous Zimbabweans working in higher education experience our own struggles when we attempt to indigenize curriculum and teaching. We went through a Western education system that transformed us into slaves of Western ideas and Western mentality. Painfully, a majority of indigenous scholars have developed into indigenous elite with political, social, and economic interests to protect. They portray themselves as purveyors of knowledge, a perception emanating from their encounter with Western education and ideology, which are supremacist. The supremacist self needs to be renegotiated in line with indigenous values and expectations that align themselves with new visions into the academy and multiknowledge production perspectives.

DECOLONIZING METHODOLOGIES

Decolonization is a process that questions colonial situations and their aftermath. Linda Smith (1999), a Maori researcher, advocates for decolonizing research projects that recover marginalized cultural knowledge, practices, and identity. Decolonizing methods are critical communication strategies that engage participants in examining lives, society, and institutions in ways that challenge dominant perspectives (Chinn, 2007). They are not only limited to research, but also to pedagogical strategies that open up avenues and spaces for critical communication. This methodology assumes that both academics or scholars and students are social actors able to engage in communicative action, defined as "that form of social interaction in which the plans of action of different actors are coordinated through an exchange of communicative acts, that is, through a use of language oriented towards reaching understanding" (Powell & Moody, 2003, p. 4). From a Habermasian perspective, communication based on participants' lifeworlds, the daily activities that make up individual existences, and intersubjective understandings of meanings establish the contexts in which personalities, society, and culture develop (Chinn, 2007). Decolonizing methodologies may be regarded as critical communication strategies that explicitly engage participants in examining lives, society, and institutions through the lenses of marginalized and dominant cultures and people.

INTEGRATING CULTURAL CONTEXT INTO RESEARCH

In Zimbabwe, research on indigenous people and their knowledge has tended to impose positivist approaches that are not holistic and inclusive.

The approach focuses on treating the researched as objects of study and not subjects and has a tendency to have the researchers' heuristic biases thrown in. According to Smith (1999),

> Some methodologies regard the values and beliefs, practices and customs of communities as "barriers" to research or as exotic customs with which researchers need to be familiar in order to carry out their work without causing offence. Research, from Smith's perspective, is bound by Western ideology; knowledge that has been colonized, and researchers that impose themselves as the "experts" who all too often perpetuate this process. Indigenous methodologies tend to approach cultural protocols, values and behaviours as an integral part of methodology. (p. 19)

Two important ways (of disseminating knowledge and of ensuring that research reaches the people who have helped make it) not always addressed by scientific research have to do with reporting back to the people and sharing knowledge. Both ways assume "a principle of reciprocity and feedback" (Mead, 1994, p. 100). Academics make assumptions about the participation of indigenous people in research and arrogantly assume in advance that people will not be interested in, or will not understand, the deeper issues. This arrogance is rooted in a colonial discursive approach with institutionalized power and privilege and the accompanying rationale for dominance in social relations (Dei, 2002). In a very real sense, research has been an encounter between the West and the "Other." Much more is known about one side of those encounters than is known about the other side. The challenge is to demystify and to decolonize these assumptions. Decolonization in research is a process of centering Indigenous experience in historical and contemporary discourse, first, by recreating and reclaiming alternative Indigenous histories and knowledges, and second, by developing Indigenous projects that contribute to the multilayered processes of self-determination (Smith, 1999).

The challenge for Indigenous academic researchers is complex. Most of the academics in Zimbabwe are and were trained in Western institutions in Europe and North America and have been indoctrinated in the use of Western research methodologies that emphasize verifiable outcomes consistent with positivism. In addition, another dimension to the complexity of the problem involves meeting the demands of academic communities based on Western traditions that have deeply ruptured the trust of Indigenous peoples in research and researchers. A key argument is that institutional structures are sanctioned by the state to serve the material, political, and ideological interests of the state and economic and social formation (Dei, 2002). Most academic researchers in African universities have internalized the colonial institutionalized power that leads them to marginalize indigenous knowledges and the participation of indigenous people in knowledge production. However, power and discourse are not possessed entirely by the power elite. Discursive agency and power to resist also reside in and among colonized groups (Bhabha, 1995). In Zimbabwe, the power shifts are witnessed in the involvement of indigenous people in conservation programs being led by the Communal Areas Management Programme for Indigenous Resources (CAMPFIRE). The program highlights that it is possible for indigenous academics or researchers to realize and recognize that indigenous people have agency and should be responsible for their social and economic resources.

COMMUNAL AREAS MANAGEMENT PROGRAMME
FOR INDIGENOUS RESOURCES

Zimbabwe's Communal Areas Management Programme for Indigenous Resources (CAMPFIRE) is widely regarded as one of Africa's most successful contemporary conservation initiatives involving indigenous people (Shizha, 2009). It permits the residents of communal lands to share in the benefits generated by wildlife utilization on their lands (Murombedzi, 1999). CAMPFIRE involves the collaboration of academic researchers and communal people in designing conservation and development goals that can be achieved by creating strong collective tenure over wildlife resources in communal lands (Shizha, 2009). From fauna and flora, communal people engage in extracting indigenous medicinal products and herbs that are important to manage indigenous peoples' health. Through CAMPFIRE, indigenous people integrate professional advice and their indigenous perspectives to living in harmony with their natural environment but benefitting from both at the same time (Murombedzi, 1999). Indigenous people use their knowledge to manage and control diseases that can affect them and their cattle. Academics assist in codifying and classifying indigenous medicinal products according to their usefulness in maintaining ecological balance. For instance, indigenous people know which type of bush and tree is important for treating different types of diseases, which trees cannot be used for firewood, and so forth. They also use their indigenous knowledge of the cosmos to plan agriculture activities and tracking animal movements.

Community knowledge and personal observation within the local environment covers pertinent angles often overlooked or neglected by the academic "objective positivist techniques." Within CAMPFIRE projects indigenous people can predict droughts as well as weather related diseases by watching the movements of celestial bodies in combination with observing the date of emergence of certain plant species which might affect their crops. Such "early warning signals" of an approaching environmental disaster are used to determine any preventive measures, prepare for mitigation, and decide on the course of the community in using the natural resources (Alexander & McGregor, 2000). Similarly, estimates of animal fertility can be drawn from such forecasts with implication on stocking rates and density. This knowledge is little used in schools and less researched so far. The VaTonga and VaKorekore in the Zambezi Valley learn names of the animals and plants, the behavioral patterns of animals as well as ecological factors under which they flourish. With the help of some indigenous researchers, indigenous people in the Zambezi valley keep inventory of species and records of those that disappear and assign names to new plants. The taxonomy reflects the use of plants for medicinal, social, economic, or cultural usefulness or other determining characteristics, as in the case of poisonous plants. Knowledge of traditional practices has not yet sufficiently been integrated into the formal educational and health domains in Zimbabwe, a missed opportunity for culturally appropriate programs.

AN INDIGENOUS LANGUAGE SCIENCE DICTIONARY

The language policy in Zimbabwe unjustifiably reproduces and perpetuates the colonial domination of the English language in education and development. Accentuating the importance of local languages in national development, Okwudishu (2006) says:

> It has been rightly observed that a national development that has not given a pride of place to indigenous languages as vehicles of national development is likely to be a wasted effort . . . development in Africa should focus on the cultivation of a literate citizenry that can participate effectively in the socio-economic, political and cultural life of the nation. Development in this sense is human-based and languages chosen for that purpose must be those that will facilitate access to information for the masses at the grassroots. (pp. 135–136)

The Zimbabwe Languages Association has described the Education Act as "characteristically colonial because it promotes English at the expense of developing indigenous languages" (Government of Zimbabwe, 1999, p. 161). Lamenting this sorry state, Sonaiya (2007) says, "what continues to be of great concern to many in Africa is the fact that even after independence not only are European languages still being maintained within the educational system, but very little is being done to develop African languages which had suffered over a century of neglect" (p. 18). This state of affairs is the most painful and absurd interface between Africa and the rest of the world and in fact Africa is the only continent in the world in which language-in-education is largely exogenous to the society it seeks to serve (Djite, 2008).

The criticism of the policy on indigenous languages assumed a high profile in 1997, when the Zimbabwe government hosted a UNESCO (United Nations Educational, Scientific and Cultural Organization) sponsored Intergovernmental Conference on Language Policies in Africa (ICLPA) that called on governments to create policies that clearly define roles for indigenous languages and the mechanisms for their development and support (Chimhundu, 2002). One of the language discourses in favor of indigenous languages in the academy has focused on indigenous languages as a resource for learning and conceptual development (Brock-Utne, 2002). Given the centrality of language to human cognition, the question arises: What should be the proper role of the academy with respect to language education in former colonial states where nonindigenous languages have remained dominant?

In Zimbabwe, English continues to dominate the academy under the pretext that English is the language of science and technology, as opposed to indigenous languages that are perceived as shallow and inadequate for use in science and technology. Rutherford and Nkopodi (1990) who adopt this perspective argue that the recognition of concepts in Western science taught in English to non-Western populations is more accessible than when taught in their mother tongue, especially when there are no linguistic correlates to the scientific concept in the indigenous language. However, Dlodlo (1999) argues that a student whose mother language has not been used in scientific discourse has additional difficulties of cognition and understanding and cannot "appeal to translation into the mother tongue for resolution of doubt or the dissipation of ignorance" (Dlodlo, 1999, p. 322). The end result is "failure" to comprehend and conceptualize "facts" on the part of the student. Contrary to the view that indigenous languages in Zimbabwe lack scientific terminology, Shizha (2008) points out that science has some biological or ecological features of the species that are reflected in the Shona lexicon. This argument is supported by the production a Shona scientific dictionary in Zimbabwe. A dictionary of biomedical terms in the Shona language titled *Duramazwi Reurapi Neutano* (Mpofu, Chimhundu, Mangoya, & Chabata, 2004) was published to reaffirm the belief that indigenous Africans languages have a scientific lexicon.

The Shona biomedical dictionary (Mpofu et al., 2004) is enriched with human biology and biomedical terms in both Shona and English, such as *tsandanyama* (muscle), *itsvo* (kidney), *chiropa* (liver), *mwoyo* (heart), *mboni* (iris), and so forth. Included in the dictionary are Shona explanations of biomedical terms such as: *chirwere chokuzarirwa* (asthma), *bhiipi* (blood pressure), *gomarara* (cancer), *chiveve* (cramp), *pfari* or *tsviyo* (epilepsy), *chirungurira* (heartburn), and many others. The dictionary has the capacity to create not only a sociolinguistic landscape, but also a major breakthrough into the scientific role of indigenous languages. This can be a major educational resource in higher education in Zimbabwe where English still perpetuates a cycle of linguistic domination and subordination of indigenous voices leading to "voicelessness" and causing "language endangerment" among indigenous students.

CONCLUSION

The educational terrain should be a site for the rediscovery and regeneration of "lost" local epistemologies. For far too long, indigenous knowledges have been neglected in the academic corridors in Zimbabwe. While credit should be given to existing projects on rethinking and reconstituting indigenous voices and knowledges in academics and research, much more still needs to be done. A generative curriculum that takes cognizance of indigenous perspectives and Western perspectives could be seen as an inclusive approach to reconstituting an academic curriculum that incorporates holism, ecological contextualism, and community-embeddedness. The modernist ideals to knowledge that depict Western modes of teaching and learning as "truth" and "one size fits all" approaches to academic pursuits are prejudicial and antithetical to understandings of knowledge as diversely determined and variably dependent on cultural settings. The language of "best practices" is foreign to an ecoculturally situated understanding of indigenous knowledge, cultural concepts, socialization practices, needs, and goals of a community. Promoting the use of indigenous languages should be sped up so that learning does not become a struggle for most Zimbabwean students who use English as a second language. Sustainable development cannot be adequately achieved in Zimbabwe if indigenous knowledges and languages continue to be marginalized and peripheralized in the academic spaces in Zimbabwe.

REFERENCES

Abdi, A. A. (2005). African philosophies of education: Counter-colonial criticisms. In A.A. Abdi & A. Cleghorn (eds.), *Issues in African education: Sociological perspectives* (pp. 2–41). New York: Palgrave Macmillan.

Adjei, P., & G. Dei. (2008). Decolonizing schooling and education in Ghana. In A. A. Abdi & S. Guo (eds.), *Education and social development: Global issues and analyses* (pp. 139–154). Rotterdam: Sense Publishers.

Aidoo, A. A. (2000). *What "hopeless continent"—The economist's perception of Africa.* Oxford: New Internationalist Publisher Ltd.

Alexander, J., & J. A. McGregor (2000). Wildlife and politics: CAMPFIRE in Zimbabwe. *Development and Change, 31*(3), 605–627.

Bhabha, H. K. (1995). *The location of culture.* London: Routledge.

Bourdieu, P., & J. C. Passeron (1979) *The inheritors: French students and their relations to culture.* Chicago: University of Chicago Press.

Brock-Utne, B. (2002). *Whose education for all? The recolonization of the African mind.* New York: Falmer Press.

Cappuccio, F. P. (2004). Commentary: Epidemiological transition, migration, and cardiovascular disease. *International Journal of Epidemiology, 33*(2), 387–388.

Chabal, P. (ed.). (1996). *The post colonial literature of Lusophone Africa.* London: C. Hurst and Company.

Chimhundu, H. (2002). *Final report: Language policies in Africa: Intergovernmental conference on language policies in Africa.* Harare: UNESCO.

Chinn, P. W. U. (2007). Decolonizing methodologies and indigenous knowledge: The role of culture, place and personal experience in professional development. *Journal of Research in Science Teaching, 44*(9), 1247–1268.

Dei, G. J. S. (2002). Rethinking the role of indigenous knowledges in the academy. *International Journal of Inclusive Education, 4*(2), 111–132.

Djite, P. G. (2008). *The sociolinguistics of development in Africa*. Clevedon: Multilingual Matters.

Dlodlo, T. S. (1999). Science nomenclature in Africa: Physics in Nguni. *Journal of Research in Science Teaching, 36*(3), 321–331.

Fanon, F. (1967). *Black skin, white masks*. New York: Grove.

Government of Zimbabwe (1999). *Report of the presidential commission of inquiry into education and training*. Harare: Government Printer.

Mazrui, A. (1993). Language and the quest for liberation in Africa: The legacy of Franz Fanon. *Third World Quarterly, 14*(2), 348–365.

Mead, A. (1994). Misappropriation of indigenous knowledge: The next wave of colonization. In *Nga Tikanga, Nga Taonga. Cultural and Intellectual Property Rights—The Rights of Indigenous Peoples*. Research Unit for Maori Education, Monograph 23, University of Auckland.

Moore, D. S. (1997). Remapping resistance: "Ground for struggle" and the politics of place. In S. Pile & M. Keith (eds.), *Geographies of resistance* (pp. 87–106). London: Routledge.

Mpofu, N., H. Chimhundu, E. Mangoya, & E. Chabata, (2004). *Duramazwi reurapi neutano* [A dictionary of medicine and wellness]. Gweru: Mambo Press (ALLEX Project, African Languages Research Institute).

Murombedzi, J. (1999). Devolution and stewardship in Zimbabwe's CAMPFIRE programme. *Journal of International Development, 11*(2), 287–293.

Nyerere, J. (1968). Education for self-reliance. In J. Nyerere (ed.), *Freedom and socialism: A selection from writings and speeches, 1965–1967*. London: Oxford University Press.

Okwudishu, A. U. (2006). Of the tongue-tied and vanishing voices: Implications for African development. In E. N. Chia (ed.), *African linguistics and the development of African communities* (pp. 129–139). Dakar: Council for the Development of Social Science Research in Africa.

Powell, J., & H. Moody (2003). The challenge of modernity: Habermas and critical theory. *Theory and Science*. Retrieved from, http://theoryandscience. icaap.org/content/vol4.1/01_powell.html

Rodney, W. (1982). *How Europe underdeveloped Africa*. Washington, D.C.: Howard University Press.

Rutherford, M., & N. Nkopodi (1990). A comparison of the recognition of some science concept definitions in English and North Sotho for second language English speakers. *International Journal of Science Education, 12*(4), 443–456.

Semali, L. M., & J. L. Kincheloe (1999). What is indigenous knowledge and why should we study it? In L. M. Semali & J. L. Kincheloe (eds.), *What is indigenous knowledge? Voices from the academy* (pp. 53–57). New York: The Falmer Press.

Shizha, E. (2005). Reclaiming our memories: The education dilemma in post-colonial African school curricula. In A. Abdi & A. Cleghorn (eds.), *Issues in African education: Sociological perspectives* (pp. 65–83). New York: Palgrave Macmillan.

———. (2006a). Continuity or discontinuity in educational equity: Contradictions in structural adjustment programmes in Zimbabwe. In A. Abdi,

K. Puplampu, G. J. S. Dei, & A. Abdi (eds.), *African education and globalization: Critical perspectives* (pp. 187–210). Lanham, MD: Lexington Books.

———. (2006b). Legitimizing indigenous knowledge in Zimbabwe: A theoretical analysis of postcolonial school knowledge and its colonial legacy. *Journal of Contemporary Issues in Education, 1*(1), 20–34.

———. (2007). Critical analysis of problems encountered in incorporating indigenous knowledge in science teaching by primary school teachers in Zimbabwe. *Alberta Journal of Educational Research, 53*(3), 302–319.

———. (2008). Globalization and indigenous knowledge: An African postcolonial theoretical analysis. In A. A. Abdi & S. Guo (eds.), *Education and social development: Global issues and analysis* (pp. 37–56). Rotterdam: Sense Publishers.

———. (2009). *Chara chimwe hachitswanyi inda*: Indigenizing science education in Zimbabwe. In D. Kapoor & S. Jordan (eds.), *International perspectives on education, PAR and social change.* New York: Palgrave Macmillan

Smith, L. T. (1999). *Decolonizing methodologies: Research and indigenous peoples.* London, UK: Zed Books.

Sonaiya, C. O. (2007). *Language matters: Exploring the dimensions of multilingualism.* Ile-Ife, Nigeria. Obafemi Awolowo University Press Limited.

Simpson, L. R. (2004). Anti-colonial strategies for the recovery and maintenance of indigenous knowledge. *American Indian Quarterly, 28*(3/4), 373–385.

Wane, N. N. (2008). Mapping the field of indigenous knowledges in anti-colonial discourse: A transformative journey in education. *Race Ethnicity and Education, 11*(2), 183–197.

wa Thiong'o, N. (1986). *Decolonizing the mind: The politics of language in African literature.* Harare: Zimbabwe Publishing House.

Wilson, W. A. (2004). Indigenous knowledge recovery is indigenous empowerment. *American Indian Quarterly, 28*(3/4), 359–372.

Zvobgo, R. J. (1996). *Transforming education: The Zimbabwean experience.* Harare: College Press.

CHAPTER 9

EDUCATION, ECONOMIC AND CULTURAL MODERNIZATION, AND THE NEWARS OF NEPAL

DEEPA SHAKYA

INTRODUCTION

The ethnic pride of the Newars, one of the ethnic communities of Nepal, is manifested in their veneration of their mother tongue, rituals, festivals, and indigenous professions (Gurung, 2000). The Newars have a well-defined occupational caste system. For centuries, they used informal settings, which included workshops, ancient manuscripts, and interactions with the elder family members to educate the young Newars on their indigenous knowledge and caste-based professions. Indigenous knowledge was also transmitted and maintained through rituals, ceremonies, and festivals. With the implementation of a centralized formal education system, the Newars have gradually been losing their informal education system. Today, formal education has become a panacea for the variety of conditions relating to individual and social advancement, such as getting better jobs or being respected by the society members. It has also been deemed important for the advancement of democracy and essential to a nation's economic and development interests. There is significant evidence that education has partially met the expectations of individual, economic, and political development (Bowers, 1997). It has brought obvious benefits, such as an improvement in literacy rates (Norberg-Hodge, 1992). While Newars suggest that formal Nepalese education—adopted from the West and based on generalizations of culture, thoughts, practices, and contents—has broadened their outlook, brought awareness about female education, provided opportunities to diversify professions, and improved their social status, paradoxically, the education system has fostered economic dependency rather than sustainable development among the Newars (Shakya, 2008). Centralized education functions as a political activity, which disconnects indigenous communities from their roots so that the communities routinely neglect their traditional practices and indigenous knowledge (Barua & Wilson, 2005). Over the years, the centralized education system has encouraged the Newars to adopt materialistic values and livelihoods at the expense of their mother tongue, indigenous knowledge, entrepreneurial skills, and sociocultural base. Because of formal education and modernization, the Newars are progressive in many aspects on the one hand, while, on the other, there has been a simultaneous and gradual marginalization of their indigenous existential situation.

This chapter discusses the situation of the Newars of Nepal in relation to the economic development and modernization agenda of the Nepali state. I critically assess the role and impact of formal education on Newari ways of life, where some changes appear to be fruitful for the community but at great cost to indigenous Newari ways. Reflections on how the Newars perceive the value of formal education (universal education) in relation to indigenous practices in the era of modernization are included, based on a qualitative case study. The author conducted the study by interviewing twelve research participants based on their consent and willingness to participate. All the participants are from the Newar community and are between the age of twenty-two and seventy-one years. The study is focused on Newari culture, modern education, and economic development and Newari community perspectives and analysis of the contradictions of the process of socioeducational change. These are pertinent considerations given the growing international interest in the role that indigenous knowledge plays in participatory approaches to development and in research on the increasing relevance of indigenous knowledge for sustainable development (Warren, Liebenstein, & Slikkerveer, 1993). This chapter highlights the initiatives of the Newars to preserve their indigeneity. The terms "ethnic," "indigenous," and "tribal"; and "modern education," "formal education," and "centralized education" are used interchangeably for convenience of discussion.

NEPAL AND THE NEWARS

Throughout history, Nepal has been known as a cultural mosaic forming a unique multiethnic, multilinguistic, and multireligious country. Prior to the unification of the geographically dispersed states and principalities into one caste, one religion, one language, one culture, and one nation in 1768, different indigenous nationalities had their own homelands and self-rule (Bhattachan, 2002). The mountain region is the ancestral homeland of indigenous ethnic groups like *Sherpas, Thakalis, Lepchas,* and *Bhotes* to name a few. Likewise, the hill regions are the home of *Tamangs, Magars, Gurungs,*

Newars, Rais, Limbus, Chepangs, Kusundas, and others. The Terai regions are dominated by groups including *Tharus, Majhis, Danuwars, Rajbansis,* and *Meches* (Bhattarai, 2006). The majority of indigenous nationalities sustained their ethnic identity, language, religion, and culture even after the unification. In Nepal, the indigenous or ethnic population is popularly known as *Janajatis.*

There is no universally accepted definition of the term "indigenous peoples," since no single definition captures the diversity of indigenous cultures, histories, and civilizations. The working definition of indigenous people proposed by the United Nations Working Group identified indigenous populations as:

> [T]hose which, having a historical continuity with pre-invasion and pre-colonial societies that developed on their territories, consider themselves distinct from other sectors of the societies now prevailing in those territories, or parts of them. They form at present non-dominant sectors of society and are determined to preserve, develop and transmit to future generations their ancestral territories, and their ethnic identity, as the basis of their continued existence as peoples, in accordance with their own cultural patterns, social institutions and legal systems.
>
> UNESCO, 2006, p. 10

Article 1 (1.b) of the International Labor Organization Convention 169, refers to indigenous people as those who inhabited the country, or a geographical region belonging to the country, at the time of conquest or colonization or the establishment of present state boundaries and who, irrespective of their legal status, retain some or all of their own social, economic, cultural and political institutions (Bhattachan, 2003). Bajracharya, Bhattachan, Dahal, and Khatry (2005) stated that indigenous nationalities possess a very precious tangible and intangible cultural heritage that has been passed from one generation to another. In the Nepalese context, the Indigenous/Nationalities Act in 2002 refers to indigenous nationalities as those groups "who have their own mother tongues and traditional customs, different cultural identities, different social structures and written or oral histories" (HMG-N, 2002, p. 170).

In the past, the Nepali government did not formally recognize the existence of indigenous people. Though their existence is accepted now, the government has yet to ratify international instruments for "indigenous nationalities," such as Convention 169 of the ILO and the United Nations Declaration on the Rights of Indigenous Peoples (Bhattachan, 2003). In 2000, the Ministry of Local Development listed sixty-one officially recognized Janajatis. Two years later in 2002, the Ministry of Law, Justice, and Parliamentary Affairs confirmed and identified only fifty-nine of them as indigenous people (Gellner, n.d.). In the National Census for 2001, the Central Bureau of Statistics (2002) listed only forty-three ethnic groups in Nepal and recorded about 37.2 percent (8.4 million) of Nepal's total population as indigenous people. The Library of Congress Country Studies (2005) mentioned that there are more than one hundred ethnic and caste groups that are classified into five larger groups on the basis of shared and prominent cultural traits: Hindus, indigenous Janajatis, Newars, Muslims, and others, a category that includes Sikhs, Bengalis, Marwaris, and Jains.

Whether the Newars should be considered as an indigenous nationality or not has been a topic of debate in Nepal. Some Newars have argued that they should not be considered as indigenous nationalities because they were something more capacious, namely, a nation. Others opined that if they are named an indigenous nationality, they will be seen as backward. So they opposed their inclusion in the official list as indigenous nationalities on the grounds that they were neither backward, nor a homogeneous cultural group. The argument raged, but in the end, the Newars remained in the list (Gellner, n.d.). There is a general belief that the Newars settled in the Nepal Valley since prehistoric times (Shrestha, 1999). The Newar kingdom made up almost the entire population of the Kathmandu Valley. The earliest known history of the Newars and the Valley was recorded in the form of several mythical scriptures. The Swayambhu Purana[1] is one such text, which accounts the creation of the Valley and settlement of the Newars. Scholars believe that the word Nepal is derived from the word *Nepa:* which refers to the Newar Kingdom. With Sanskritization,[2] the Newar word *Nepa:* became Nepal, representing the name of the unified country.

The Newars make 44 percent of the total population of the Valley; however, nationally they make about 5.6 percent of the total population (Gellner, 1995). There are Buddhists as well as Hindus among the Newars, who are bonded together by their mother tongue, Nepal Bhasa.[3] They are the inheritors of rich history, culture, and indigenous skills. They are prominent in every sphere, from agriculture, business, education, and government administration to medicine, law, religion, architecture, fine art, and literature. The Newar architecture, sculptures, wood and stone cravings, and metal craftsmanship are world-renowned for their exquisite beauty. The fine temples and palaces of Kathmandu, Patan, and Bhaktapur are largely the product of Newar architects, artisans, and sculptors. The Newar historical and architectural contributions to the Valley were included in the UNESCO world heritage list in 1979 (Bhattarai, 2006).

The Newars maintain a highly sophisticated caste system. Caste has been known as an element in the social structure of the Nepal valley since the Licchavi[4] period, and the Newar caste system has had its own development apart from the other caste systems in Nepal. Various historical sources show that the Newar caste system was first codified by Jaya Sthithi Malla[5] in the fourteenth century (Löwdin, n.d.). The Newar caste system has several layers, which comprise more than forty distinct hierarchical subgroups representing different occupational caste, readily identified by surnames (Bhattarai, 2006). For example: Rajopadhyas are the priests for all Hindu Newars; Vajracharyas are the priests for all Buddhist Newars; Shakyas are the temple priests and also traditionally gold and silversmiths; Joshis are the astrologers; Karmacharyas are the tantric priests; Shresthas are mainly traders; Tuladhars and Baniyas are merchants; Vaidyas are ayurvedic practitioners, Sikarmis are carpenters; Tamrakars are metal smiths; Lohnkarmis are stone workers; Kansakars are workers in bronze; and Maharjans/Dangols are farmers (Gellner, 1995). Generation after generation, the Newars grew up learning and implementing their indigenous practices, which helped the Newar community gain the reputation of being an independent community in the country. Each caste group performed the caste-based occupation and things went on smoothly without any qualms.

The indigenous knowledge, embedded in the Newar community in the form of folk literature, dance, music, art, artifacts; clothing and housing; fairs and festivals; life cycle rituals; and traditional healing practices, has been developed outside the formal educational system (Bajracharya, Bhattachan, Dahal, & Khatry, 2005; Guchteneire, Krukkert, & Liebensteinet, 2003). Various informal mechanisms were used to transmit knowledge from one generation to another. Traditional workshops (the Newars refer to them as *Jyasaw*) applied the "learning by doing" approach to knowledge transfer, especially with respect to arts and crafts. Priests, indigenous medical practitioners, astrologers, and the like mostly referred to ancient manuscripts and practices in one-on-one or group teaching sessions. Knowledge gained through these vehicles was then assimilated and used in everyday living. The youngsters acquired indigenous knowledge through conversations and interactions with their elders. Children were sent to temples (Hindus) and monasteries (Buddhists), which played important roles in enhancing moral values and knowledge of history and the mother tongue. Indigenous knowledge was also transmitted and maintained through ritual, ceremony, and art. The *guthi*[6] indirectly contributes in the transmission of Newari culture and knowledge from one generation to the next. It also plays an important role in preserving religious, historical and cultural activities; maintaining a harmony in the Newari society; and in preventing anarchy (Diwasa, Bandhu, & Nepal, 2009). The Newars' indigenous education system was gradually marginalized with the implementation of centralized education in Nepal. They came to feel that essential knowledge came from course books and formal education. While the Newars flourished in several aspects with formal education, they overlooked their indigeneity in the name of development and modernization.

ECONOMIC AND CULTURAL MODERNIZATION OF THE NEWARS: ROLES AND IMPACTS OF FORMAL EDUCATION

Economic growth is an essential factor for climbing out of poverty and pacing up in the global race for a developing country like Nepal. Education is not only widely accepted as a leading instrument for promoting economic growth (Bloom, Canning, & Chan, 2005), but also is seen as essential to a nation's development interests. After the overthrow of the Rana rule and the advent of democracy in 1951, Nepal began its drive for socioeconomic development. The government then opened schools for the general public and adopted formal education as the backbone of the country's development. In the Secondary Curriculum Introductory Training Booklet (1999), the Curriculum Development Centre stated:

> Education plays a very important role in the development of the individual personality, society and the nation. It plays a vital role in broadening the people's vision. It is said that education is the light of life. For the all round development of the country, human resource development is a must. It is education which helps to produce national heroes, disciplined citizens, industrious manpower and able and suitable human beings for society. It is said that educated skilled human resources are the backbone of the nation. In their absence, a country cannot progress. (p. 11)

Education in Nepal is divided into three phases: primary, secondary, and higher education. Primary education covers five years (Grades 1–5). Secondary education comprises three cycles: three years of lower secondary (Grades 6–8), two years of upper secondary (Grades 9–10), and two years of higher secondary education (Grades 11–12). A national-level School Leaving Certificate (SLC) Examination is conducted at the end of Grade 10. After passing the SLC level, students become eligible to purse their higher secondary education in any of the four streams: science, commerce, humanities, and education (World Data on Education [WDE], 2006/07). At the end of these studies, students sit for the Higher Secondary Certificate supervised by the Higher Secondary Education Board (HSEB). The successful completion of this opens doors for students to pursue higher education: bachelors, masters, and PhD levels (Bajracharya, Bhuju, & Pokhrel, 2006). The medium of education is primarily English in private institutions and a mix of English and Nepali in public institutions. Naresh, a Newar who has become a successful professional consultant and strategic marketer said:

> Formal education, for me, is foreign. Yes, science and mathematics along with other subjects do tend to uplift society and community but not all subjects. With the positives that education brings that is beneficial; it brings a lot of negatives as well. Most of the negativities result in alienation of the Newars from their culture.
>
> Shakya, 2008, p. 58

Centralized education was not a part of the Newars' habitual indigenous lifestyle. Among the elder generation there were only a handful of Newars pursuing formal education. Since the people were leading a sustainable life, they rarely required any advanced educational expertise and degrees to perform the indigenous functions. However, centralized education has not only been presented as the right of all children, but a passport to a better future. The Newars started to enroll their children in schools hoping for their betterment. Also the community began to look at the educated with more dignity and respect. The higher caste Newars initiated the trend of pursuing formal education and the lower castes gradually followed it. As Rahul, one man in my study reported: "The Newars are now aware of the importance of education and almost each person of our society is conscious about educating their children for better status and living standards" (Shakya, 2008, p. 57).

According to a Nepal South Asian Center's (NESAC) report, the Newars are ranked first out of one hundred ethnic/caste groups in Nepal in terms of longevity, knowledge, and standard of living. The per capita income of the Newars was NRs11,953 in 1998 according to the NESAC records (Bajracharya, Bhattachan, Dahal, & Khatry, 2005), which is the highest among the ethnic/caste groups of Nepal. Among other factors, the Newars attribute their development to their acquaintance with formal education. In an interview, Rahul highlighted the benefits of formal education and said:

> Education is the base for living a better life. Educated society is always responsible for the betterment of society. So education has played a vital role in maintaining a good social environment. The education I got has also helped me a lot in handling my day-to-day activities and problems. Better quality education has helped in getting better jobs.
>
> Shakya, 2008, p. 56

Today, one will find among the Newars not only farmers, traders, craftsmen, or low level workers but also scores of distinguished doctors, engineers, professors, respectable government personnel, service holders, advocates, businessmen and the like. This proved to be beneficial for many Newars, especially the lower castes as it raised their status and standard of living. In 2004, the literacy rate of the Newars was higher than the national average of 53.7 percent (UNDP, 2004). Over the years, the Government of Nepal has implemented various awareness programs on formal education, stressing more on the education of girls. Consequently, there has also been an increase in schooling of the Newar girls. Previously, only males were sent to schools and had economic freedom. Females were basically restricted to housework. Formal education opened avenues for the female Newar population to be economically independent. Today, we can find many professionally and economically independent Newari women involved in different economic and development sectors as opposed to just being housewives or complementing their family's or in-law's work. Bimala, who partners her husband's business said:

> Education has had positive effects on the social environment. It has made our society educated and well organized. In addition, it is due to the influence of education that Newari women have been able to work outside the house. Formal education has enabled me to equally partner my husband in the sweetmeat business.
>
> Shakya, 2008, p. 72

The educated Newari women have contributed in lifting the economic activity and improving the health and hygiene of the Newari people. If measured in terms of indicators, such as per capita income, life expectancy, literacy, levels of employment, and the human development index, we can say that there has been socioeconomic development among the Newars. From an optimistic view, modern education gave the Newars a broad outlook of life and an undaunted zeal to upgrade their standard of living as well as a means of better livelihood. It helped them become participants in the ever-changing global race. But, the escalation of the Newars toward lucrative market opportunities has resulted in the dearth of hierarchy-based professionals. For example, since each caste had to fulfill its own duties in the festival and rituals, transformation of socioeconomic life has paralyzed many rituals and festivals and caused despair among the Newars. In this context Naresh said:

> The Newars have themselves to blame. There is no concerted effort to diffuse modern education into our indigenous system. The mix of indigenous knowledge and universal education is still like oil and water. There is polarization and as our wealth creation methods cease to provide sustenance, opting for employment becomes normal. While service is not bad, it has not been ideally ingrained into our system. How many Newars have actually made it to the top?
>
> Shakya, 2008, p. 77

Furthermore, there are a number of Newars who graduated from college, but failed to find place in the job market. These educated Newars carried the tag of an educated person, and it was against their dignity to go back to the ancestral profession. Among these, only a handful of them are seen continuing the indigenous profession, while the majority of them preferred to stay idle or take underqualified jobs. This situation exists because universal education acts almost as a blindfold, preventing the students from seeing their indigenous contexts and, worse, to look down on them. Norberg-Hogde (1992) stated that everything in schools promotes the Western model and as a direct consequence makes students ashamed of their own indigenous systems. In the process of neglecting their own self-sufficient economic strength, the Newars have fallen into the dependency economy syndrome. Naresh said:

> On a macro level, it is interesting to note that Newars are a clan that had socio-economic classifications. I belong to the skill crafts people/business class. However, with education, we learnt the need to have a career, a job, an employee attitude, secure life and wealth distributor while we ceased to become a business person and a wealth creator. When I look at my cousins who were not so fortunate with education and feel they are living in the past, being conservative ignorant of the new world, I feel lucky. When I look at myself compared to how successful they (goldsmiths) are financially, having time to make merry and involve themselves in all social activities I feel I have underachieved. I always believed and realized that formal education taught us

to be followers and never leaders. I am not sure, which education to compare but it seems our forefathers were wealthier, leading a sustainable life and content. While at the same time, present day education has prepared us to meet the challenges of this very dynamic and dependent this society has become.

                                                            Shakya, 2008, p. 75

The elderly Newars were not equipped with formal education, yet they were successful in establishing a systematic and sustainable lifestyle. They inherited their ancestral caste-based professions. It made them proud of their culture and they tended to think of themselves as being the true culture bearers of Nepal. Today, the Newars have become more fixated on pursuing degrees, getting hold of jobs, earning money, and encouraging self-centered development. The pursuit of knowledge, social engagement, participation, and critical thinking, which prevailed in indigenous settings, have become a thing of minor importance (Barua, 2007), while imposed class lectures and theories, and degrees and certificates gained major importance in the formal education system.

The centralized form of formal education is based on generalizations of cultures, practices, languages, and knowledge. In Nepal, the curriculum designed under the formal education system primarily focuses on promoting values of the ruling authority in the name of national culture, and creating human resources to supplement national development (Niroula, 2007). Indigenous knowledge is excluded and marginalized in the process of formal education and development (Barua, 2004). The indigeneity of ethnic groups like the Newars has been put at stake with the implementation of a centralized curriculum as opposed to a curriculum that addresses cultural diversity. There is an absence of classroom discussion on history, civilization, indigenous technology, and culture of the diverse ethnic groups.

In an interview, Harka, a seventy-one-year-old Newar with a PhD degree in Linguistics, said, "Our tradition and culture have not been promoted in any significant way and mother tongue education has been neglected at the starting point of education" (Shakya, 2008, p. 59). The Valley used to be full of hymns and conversations in Nepal Bhasa. Now, we hear more of Nepali and English. A trend of speaking Nepali in Newar houses basically started during the Rana period. First, it was adopted by members of the Newar elites who used to serve the Ranas. When the government declared Nepali as the national and official language, government aided schools used Nepali as the medium of instruction. Thus, high- and middle-class Newar parents were more inclined to speak Nepali with their children at the cost of their own mother tongue to ensure their children's success in schools. The Newars also had to forsake their language in order to obtain better job opportunities in offices. The job opportunities always have criteria, "should be fluent in spoken and written Nepali." With English boarding schools mushrooming in the Valley, the younger generations of Newars are busy equipping themselves with both Nepali and English languages. Today, most of the younger Newars are totally alien to their own mother tongue. While a few can speak and understand, reading and writing in Nepal Bhasa script is something that only the elders treasure. Hira, a young Newar boy and a struggling journalist said:

> If we take culture as the set of values and traditions, and so on, then I must say that our culture is degrading. Our language will soon be disappearing. Our culture is vaguely misinterpreted and dyed with mysticism. Our very origin and the Newar stratification are complex and different sources tell different stories, which are controversial.
>
> Shakya, 2008, p. 61

Formal education familiarized learners with the concept of modernization and encouraged them to adopt the so-called modern lifestyle. Modernization has labeled people as "developed" or "underdeveloped" in terms of per capita income, share of GDP, and literacy status (Hoogvelt, 1982). The Newars, fascinated by the "developed" label, welcomed modern luxuries like audio/video, private vehicles, concrete buildings, and other modern amenities into their lifestyle. Modernization eased the daily routine of the Newars and brought efficiency to their occupations, while it has also been seen as a threat to the indigenous way of life. Nirmala, who assists her husband in his traditional Newar metal craft profession, said:

> I regard modernization to be very important in the enhancement of the occupation I am involved in. I assist my husband in his metal work and because of modernization we are equipped with different modern technology and equipments that have helped us to improve the quality of our metal products and made the whole process efficient.
>
> Shakya, 2008, p. 62

With the growing pace of modernization, the Newar families started to prefer living in nuclear families where, in most cases, both parents are occupied in their employment. They are left with little or no time to cherish their culture. In an interview, Arjun, who is a goldsmith by profession and currently pursing his bachelor degree in the field of arts, said:

> I have seen that the Newari culture is only remaining in joint families. Once these families are split into several nuclear families they ignore it in the name of modernization. The new generations are more attracted towards Western culture and Indian cinema than in their own music and culture. Except for few elders in less urbanized corners of Patan and Bhaktapur cities, no one wears traditional Newari clothes like *haku patashi* or *Lann Surwal*. Instead, they wear fashionable Western clothes. No one has time to celebrate their festivals; instead Western festivals are celebrated more and given preference.
>
> Shakya, 2008, p. 63

Increasingly, the Newars are becoming attracted to new trends of thought and development. The worry here is that the globalization process facilitated by the Western educational system is systematically universalizing the world knowledge system and weeding out all other forms of knowledge systems, institutions, and resources that are not Western in origin (Guri, 2007). Shrestha (1999) expressed his fears that these growing intrusions of foreign culture into the Newar society might, in the long run, eliminate the pride and identity of indigenous Newars.

NEWAR RESPONSES AND PROSPECTS:
SOME CONCLUDING REFLECTIONS

The basic component of any country's knowledge system is its indigenous knowledge, which evolves around the skill sets, experiences, and insights of people, and applied to maintain or improve their livelihood (World Bank, 1997). The dominance of the Western knowledge system aided by development intrusions has largely led to a situation where indigenous knowledge is neglected and ignored. Indigenous practices are fading away as they seem to become inappropriate or too slow to meet new challenges. This disappearing act of indigenous knowledge not only impacts the ones who carried it also causes permanent loss or disappearance of skills, technologies, artifacts, problem solving strategies, and expertise all at once.

In order to survive in the modernized global economy, the Newars have embraced new ways of living. They remain occupied in competing to excel in the complex and dynamic society of today and that hardly allows them to foster their cultural and traditional values. In an interview, Biren who is a lecturer by profession said, "Most people of the Newar community, even the educated ones, want to continue their traditions, if possible. But, the lack of time is an impediment to keep up the tradition and culture" (Shakya, 2008, p. 61). If we study the recent trend among the Newars, we will hardly find any highly educated Newars continuing their indigenous professions. Those who have continued the professions have less educational background. A few educated Newars are successful in moving ahead and meeting modern demands, while there are many trapped in a dilemma. They are neither able to excel in the modern economy nor able to go back to their indigenous professions. This calls for a mechanism to bind formal education and indigenous knowledge together through collective action, so that the Newars do not find themselves inclined to only one aspect while being alien to another or confused about both.

The UNESCO-ICSU World Conference on Science, in its 1997 declaration, stated: "traditional and local knowledge systems, as dynamic expressions of perceiving and understanding the world, can make, and historically have made, a valuable contribution to science and technology, and that there is a need to preserve, protect, research and promote this cultural heritage and empirical knowledge" (UNESCO, 1999). Newari culture, arts, and artifacts are widely regarded as important resources for the country's economy and cultural tourism. The Newars feel that their culture, arts, and artifacts need to be promoted to preserve their identity and indigeneity from becoming museum artifacts. James D. Wolfensohn, President of the World Bank, said that "indigenous knowledge is an integral part of the culture and history of a local community," and that "we need to learn from local communities to enrich the development process" (Cited in Gorjestani, n.d., p. 1). Thus, the indigenous communities come with a shared realization that they need to enter the new millennium facing the challenge of reinventing almost everything many of them have taken for granted (Korten, 1999).

Positive changes are taking place in the Newar society to preserve their indigeneity but it is happening in small circles. A few Newar groups have been teaching children and youth to play the indigenous instruments and participate in various cultural events. The young ones are involved in these as a part of their hobby or interest. However, until these interests are given proper recognition and respect by the society and the nation as a whole, its continuity is hazy. The modernized way of learning and living has to see the value in indigenous practices so that the younger generations continue to bind these to their present way of life. In order to promote the knowledge and use of Nepal Bhasa, two Nepal Bhasa dailies *Sandhya Times* and *Visvabhummi* are in circulation, but they are read by only those able to read the Nepal Bhasa script. Similarly, various national TV channels and FM stations broadcast programs and news in Nepal Bhasa. But these have not been very effective, as only a selected few watch and listen to these, while the rest are attracted to the Western and Indian channels/programs. Greater effort, therefore, should be undertaken to strengthen the capacity of local people to develop their own knowledge base and to develop methodologies to promote activities at the interface of scientific disciplines and indigenous knowledge (Warren, Liebenstein, & Slikkerveer, 1993). The Newars need to put in a strong effort not only preserve their indigeneity but to make it vibrant. This would only be possible if the Newars come forward for collective action to voice the need for integrating their rich heritage into the educational curricula, media, and development programs. Without collaborated efforts, not only the cultural practices and identities of ethnic groups, but generations of civilization and heritage will become extinct.

REFERENCES

Bajracharya, R. D., K. B. Bhattachan, D. R. Dahal, & P. K. Khatry (2005). *Cultural and religious diversity: Dialogue and development.* Kathmandu, Nepal: United Nations Educational, Scientific and Cultural Organization (UNESCO).

Bajracharya, D., D. R. Bhuju, & J. R. Pokhrel (2006). *Science, research and technology in Nepal.* UNESCO Kathmandu Series of Monographs and Working Papers, 10. Kathmandu: UNESCO. Retrieved from, http://unesdoc.unesco.org/images/0014/001461/146117e.pdf

Barua, B. (2004). *Western education and modernization in a Buddhist village of Bangladesh: A case study of the Barua community.* Unpublished doctoral dissertation, University of Toronto, Canada.

Barua, B. (2007). Colonialism, education and rural Buddhist communities in Bangladesh. *Journal of International Education, 37*(1), 60–76.

Barua, B. P., & M. Wilson (2005). Agroforestry and development: Displacement of Buddhist values in Bangladesh. *Canadian Journal of Development Studies, 26*(2), 233–245.

Bhattachan, K. B. (2002). *Traditional local governance in Nepal.* Seminar Paper, Political Science Association of Nepal (POLSAN) and Friedrich-Ebert-Stiftung (FES), Kathmandu, Nepal, April 21, 2002.

Bhattachan, K. B. (2003). *Indigenous nationalities and minorities of Nepal.* Minority Rights Group International Report, London, United Kingdom.

Bhattarai, S. (2006). *The Bola or Parma of the Newar in Manamaiju Village: The significance of a farm labor exchange system among indigenous peasants in Nepal.* Unpublished master's thesis, University of Tromsø, Norway. Retrieved from, http://www.ub.uit.no/munin/bitstream/10037/154/1/thesis.pdf

Bloom, D., D. Canning, & K. Chan (2005). *Higher education and economic development in Africa.* Retrieved from, http://siteresources.worldbank.org/EDUCATION/Resources/278200-1099079877269/547664-1099079956815/HigherEd_Econ_Growth_Africa.pdf

Bowers, C. A. (1997). *The culture of denial: Why the environmental movement needs a strategy for reforming universities and public schools.* Albany, NY: State University of New York.

Diwasa, T., C. M. Bandhu, & B. Nepal (2007). *The intangible cultural heritage of Nepal: Future directions.* Kathmandu: The United Nations Educational, Scientific and Cultural Organization, Kathmandu Office.

Gellner, D. N. (1995). *Contested hierarchies: A collaborative ethnography of caste in the Kathmandu Valley, Nepal.* Oxford: Clarendon Press.

Gellner, D. N. (n.d.). *Newars as Janajatis.* University of Oxford. Retrieved from, http://www.isca.ox.ac.uk/staff/academic/gellner/documents/NewarsasJanajatis.pdf

Gorjestani, N. (n.d.). *Indigenous knowledge for development: Opportunities and challenges.* Retrieved from, http://www.worldbank.org/afr/ik/ikpaper_0102.pdf.

Guchteneire, P., I. Krukkert, & G. Liebensteinet, eds. (1999). *Best practices on indigenous knowledge.* The Hague and Paris: The Management of Social Transformations Program (MOST) and the Centre for International Research and Advisory Networks (CIRAN).

Guri, B. Y. (2007). *Innovation approached to international education—Indigenous knowledge and human capital formation for balanced development.* Retrieved from, http://www.compasnet.org/afbeeldingen/2007-11-27%20Article%20Bern%20Guri%20%20Innovative%20Approach%20to%20International%20Education.pdf

Gurung, P. (2000). *Bungamati: The life world of a Newar community explored through the natural and social life of water.* Unpublished master's Thesis, University of Bergen, Norway.

Korten, D. C. (1999). *The post-corporate world: Life after capitalism.* West Hartford, CT and San Francisco, CA: Kumarian Press, Inc. and Berrett-Koehler Publishers, Inc.

HMG-N, Central Bureau of Statistics (2002). *Population census 2001.* National Report, Kathmandu: HMG-N, National Planning Commission Secretariat, Central Bureau of Statistics in collaboration with UNFPA.

Hoogvelt, A. M. M. (1982). *The third world in the global development.* London: Macmillan Press.

Library of Congress Country Studies. (2005). *Country profile: Nepal.* Retrieved from, http://lcweb2.loc.gov/frd/cs/profiles/Nepal.pdf

Löwdin, P. (n.d.). Food, ritual and society: A study of social structure and food symbolism among the Newars. Unpublished doctoral dissertation, University of Uppsala, Sweden.

Niroula, S. P. (2007). Political change and education reform in Nepal. In *Educational systems issues around the world* (pp. 34–36). Costa Rica: University for Peace. Retrieved from, http://74.125.93.132/search?q=cache:3GBgCa_qf4kJ:www.upeace.org/about/Newsflash/4_6_07/07-05-14%2520Educational%2520Systems%2520Issues%2520Around%2520the%2520World.pdf+Educational+systems+issues+around+the+world&cd=2&hl=ne&ct=clnk&gl=np&client=firefox-a

Norberg-Hodge, H. (1992). *Ancient futures: Learning from Ladakh.* San Francisco, CA: Sierra Club Books Paperback Edition.

Secondary Curriculum Introductory Training Booklet (1999). Bhaktapur, Nepal: Curriculum Development Centre.

Shakya, D. (2008). *Universal education, culture and socio-economic development in the Kathmandu Valley: A case of the Newars in Nepal.* Unpublished master's thesis, East West University, Bangladesh.

Shrestha, B. (1999). Newars: The indigenous population of the Kathmandu Valley in the modern state of Nepal. *Contributions to Nepalese Studies, Centre for Nepal and Asian Studies (CNAS) Journal, 26*(1), 83–117.

United Nations Development Program (UNDP) (2004). *Nepal human development report 2004.* Kathmandu: Author. Retrieved from, http://www.undp.org.np/publication/html/nhdr2004/NHDR2004.pdf

United Nations Educational, Scientific and Cultural Organization (UNESCO) (1999). *Declaration on science and the use of scientific knowledge.* World Conference on Science. Retrieved from, http://www.unesco.org/science/wcs/eng/declaration_e.htm

――――. (2006). *UNESCO and indigenous peoples: Partnership to promote cultural diversity.* France: Author. Retrieved from, http://unesdoc.unesco.org/images/0013/001356/135656M.pdf

Warren, D. M., G. W. Liebenstein, & L. J. Slikkerveer (1993). Networking for indigenous knowledge. *Indigenous Knowledge and Development Monitor, 1*(1), 2–4.

World Bank (1997). *Knowledge and skills for the information age: The first meeting of the Mediterranean development forum.* Retrieved from, http://www.worldbank.org/html/fpd/technet/mdf/objectiv.htm

World Data on Education (WDE) (2006/07). Retrieved from, UNESCO International Bureau of Education website www.ibe.unesco.org/countries/country-Dossier/natrep96/nepal96.pdf

PART III

LEARNING AND COMMUNICATIVE MEDIUMS

CLASH OF ORALITIES AND TEXTUALITIES: THE COLONIZATION OF THE COMMUNICATIVE SPACE IN SUB-SAHARAN AFRICA

ALI A. ABDI

INTRODUCTION

Before the arrival of colonialism, most societies in sub-Saharan Africa were oral communities whose languages were not written. This is interesting because Africa was the first place in the world where fully written linguistic characters were established and used for communication with the introduction of the hieroglyphics alphabet and structure of writing in ancient Egypt in about 3000 B.C. The Egyptian system represented a system of writing that mainly contained pictorial characters. As such, it was sometimes called the picture script of ancient Egyptian priesthood.

The major breakthrough in transforming the spoken word into something that was legible took place (slowly and in a transitional way), "when a coded system of visible marks was invented [and used] whereby a writer could determine the exact words that the reader would generate from the text" (Ong, 1985, p. 84). For some monocentric historians, the power of the written word may have moved the world into new civilizing possibilities. This may have also heralded the beginning, as colonialism definitely intended, of the philosophico-epistemological onslaught (by textual societies) on both the psychosomatic and pedagogical existences of oral communities.

In reality, though, the argument that the invention of writing represented the contextual validity of history should be a misplaced one, for the so-called prehistory (before written history) was anything but void of human ingenuity, inventiveness, and development.

For me at least, this prehistory was full of active citizenships that used powerful schemes of oral communication to design life, modify contexts of living, and precisely predict events and future conjectures of the cosmos. Needless to say, the hegemonic formations of all languages, either through their written status or otherwise, usually follow a political trajectory that elevates the power of one or more languages at the expense of all others (Bourdieu, 1991). As such, and contrary to the colonial project that either destroyed African languages and learning systems or relegated them to a status of perforce established untenability (Achebe, 1996[1958]; Nyerere, 1968; Rodney, 1982), orality was not a linguistic deficiency but rich and original ways of reading and conveying the world. Even Ferdinand de Saussure, dubbed the father of modern linguistics, acknowledged the primacy of oral speech, as representing the original physics as well as the inherent consciousness of the primordial use of languages. As such, one might say that written communication should be an additive to orality, rather than seen as replacing or reforming it.

In the way I am using the term "communicative" in the title of this chapter, it represents all linguistic or sign-based exchanges that take place among individuals, within groups, and between societies. Indeed, every utterance has a communicative message; as such, all languages and their signs and symbols are methods and modes of communication that give meaning, explain contexts, and command or sanction the practice of the communicated message. In addition, orality is critically situated against the expansive and now across-the-board generalized notions and actions of textuality, which does not only speak about the specific and almost always measurable qualities of written texts, but historically took the added connotations of strictly adhering to the contents and the fundamental interpretations (i.e., interpreted and socially located as was originally intended) of what was bound within the confines of the text. In contrasting these two communicative realities, the chapter advances the richness of oral traditions, and how they have been trumped by the imposition of textual forces that, on the basis of selfish colonial interests, were bent on de-epsitemologizing and by extension, deontologizing the historico-culturally located primary communicative lives of subordinated populations.

ORALITIES AND TEXUALITIES: DIFFERENT WAYS OF RELATING TO THE WORLD

At more psychocultural and ontological levels, it may useful to argue that the different effects orality or the written text have on its users could shape the humanistic dispositions that influence the way we relate to others. Orality brings people together, its form of communication assures more physical presence among different interlocutors, and one's opportunity to explain self, situation, and needs is close, visible, and tangible. The written text, on the other hand, is generally depersonalized, detached, more often desubjecti- fied, and mechanically constituted. Indeed, the communication relationship is mostly among pages of papers and individuals, many of whom may not be endowed to effectively respond to the commands and instructions of the often alienating text. As such, oral societies are more inclusive in the ways they relate to other human beings, more willing to understand the needs of the Other, more open to the ongoing exchanges that affirm the continuum of learning, more attached to the emotions and subjectivities of the speaker, more capable of immediately seeing where they are misconnected from the intentions of their discussant, more situated to quickly respond to emerg- ing interpersonal needs, and more attuned to the possibilities of colearning. In all, oral societies are more linguistically and communicatively attached to the social context, and thus endowed with more human qualities that counter objectification, and in the general sociophysical environment, are more located and ontologically affirmed within and around morally binding existentialities that refuse to oppress and marginalize others.

Addressing some of these fundamentally unique aspects of oral traditions, Sapir (cited in Anderson, 1988) pointed out how "many [so-called] primitive languages have a formal richness, a latent luxuriance of expression, that eclipses anything known to the languages of those who claimed to have been the civilized nations. Popular statements as to the extreme poverty of expression to which primitive languages are doomed to, are, therefore, simply myths" (p. 76). Indeed, as JanMohamed (1983) noted, the medium of the written word eventu- ally "destroys the immediacy of personal experience and the deeper socialization of the world, and consequently the totalizing nature of oral cultures" (p. 280). That is, more or less, to say, that because oralities as media of communication and expression are so close to the reality of the implicated life situations, they are concretely more real, and above all else, more open-ended in time and space. Ong's (1985, partially citing Havelock, 1963) comments are useful here:

> In the absence of an elaborate analytic categories that depend on writing to structure knowledge at a distance from lived experience, oral cultures must conceptualize and verbalize all their knowledge with more or less close refer- ence to the human life world assimilating the alien, objective world to the more immediate familiar interaction of human beings. (p. 42)

Indeed, while we are not claiming that the humanized subjectivity of orality vis-à-vis the detached objectively calculating nature of textuality should fully explain the outcomes of the colonial encounter, one might still safely argue that the subjective nature of Africa's orality was important in fully seeing the humanity of invading Europeans, which may have prematurely instilled in the former a sense of trust including the promise that their lands and people would not be oppressed and exploited. It was not actually a specific socio-historical naivety specific to Africans or colonized populations that triggered the issues; it was an important component of experience being deployed as knowledge, an issue that is now established and accepted in contemporary theories of knowledge. In simple terms, if you have not seen or practiced colonization, you will only believe it when you see it. Because of the subjective nature of the orally based linguistic dispositions, the African worldview and culture which were communal with a noncompetitive, cooperative ownership of land and other valuable resources were extensively attached to the indigenous philosophy of *Ubuntu*. *Ubuntu* is laden with extensive social, ethical, and value implications, and prioritizes the full recognition of the humanity as well as the rights of the Other. Krog (1998) situationally and rather holistically defines *Ubuntu* as representing "an African philosophy of humanism that emphasizes the link between the individual and the collective" (p. 373), or as Swanson (2007) locates it, the belief that "a person is a person through their relationship to others" (p. 55). The opposite to the "ubuntuizations" of life was forthcoming from the presumed European center: the extensively otherized Africans were portrayed as not only different but unequal, uncivilized, and negatively educated (superstitions, witchcraft, infantile magic, etc.), tribalistic, and lacking any morals or high culture, and therefore (went the hastened justification) fit to be colonized, and by extension, destroyed at will (Hegel, 1965; Said, 1993). In such a clash of ideas, ways of knowing, communicating, and living, those who wanted to humanize would immediately lose to those who were effectively prepared to dehumanize. It is via this dehumanization, and with the critical understanding that to subjugate and exploit people you need to successfully rescind their histories, cultures, and ways of knowing and communicating, that oralities were derided as naturally inferior to textualities, and relegated to a status of subordination and ridicule. As already implicated above, though, oral communication is an effective and powerful way of living, connecting, and successfully developing and managing complex life systems that expansively sustain and modify the lived contexts of people.

With these *a priori* assumptions about Africans and African life, the mostly oral traditions of these societies' languages were neither being appreciated nor promoted as media of communication, means of education, or otherwise. More problematically, African oral literatures were not accepted as genuine valid forms of social, cultural, political, legal, and economic expression. For the European powers, only written literature (in colonial languages) was to be regarded as meaningful literature. As Ngugi wa Thiongo (1998) noted, "the privileging of the written over the oral had roots in the relationship of power in society and history . . . The dominant social forces had become identified with the civilized and the written" (p. 108).

ORALITY AND ITS METHODS: THE NARRATIVE, THE STORY, AND THE PROVERB

That, of course, should not be the case, for unwritten languages used by tradi-tional, oral societies have had valid and time-tested values for learning about and managing community affairs as well as sanctioning unwanted practices, responsibly exploiting environmental resources, and even creating special-ized literary subcultures within one main culture. As Andrzejewski (1985) pointed out, "there is a mounting evidence that the presence of literature is a universal characteristic of human society; literature is an art which uses language as its medium irrespective of whether it is oral or written" (p. 31). Indeed, the diversity, with hierarchical intentions, of the communicative methods (the narrative, the story, proverbs, and poetry) used by oral societies affirms the sophisticated literature based nature they applied to different con-texts of life. While poetry (discussed in the following section) may have been regarded as verbally the most organized and as such the most socially valued way of communication in oral societies, the other methods were also power-ful conveyors of the important issues that communities wanted to deal with. In terms of the narrative, for example, Ong (1985) noted that "although it is found in all cultures, narrative is in certain ways, more widely functional in primary oral cultures than in others . . . most, if not all, oral cultures gener-ate quite substantial narratives or a series of narratives" (p. 140). And while the use of narrative was actually banished in the now extensively textualized systems of education for a long time, the return of the narrative methodol-ogy in the past twenty or so years as a legitimate and in some cases desirable platform of knowledge analysis and creation, testifies to perhaps our gradual realization that the more we subjectively observe and explain social contexts and relationships, the better we understand those contexts. As Kaplan (2003) noted (relying on select Ricoeurian analysis and affirming the intersubjective nature of narrative communication), "narrative discourse configures human actions, already prefigured like narrative, into a coherent whole that is then refigured by the [contemporary] reader" (p. 47). In storytelling, a widely used method of education in oral societies, famous storytellers are, in many cases, as respected and referred to as poets. In many African societies, as Pellowski (1977) says, "there is still a high priority assigned to family storytelling . . . Children of the Ewe people of Ghana are simply not considered educated unless they have heard many times the *gliwo*, animal stories . . . that teach about different aspects of life" (pp. 44–45).

It is possible that in many cases, storytelling will come naturally to some, but in most cases it is a method that is learned (Fadiman, 1994). In storytelling, especially in oral societies where it is a prominent process of education, the relationship between the speaker and the listener must be permanent, selectively reinforced, and continually monitored. As Brown (1995) observed, "one of the sources of risk in communication is that

whereas speakers may think that what they have to say is sufficiently important to be paid attention to, listeners (even younger ones) may have other
priorities and may not listen in detail but only partially, or perhaps not at all"
(p. 26). So what Habermas (1998) calls *universal pragmatics* in communication, that is, a reliable validity of understanding between the speaker and the
listener is "technically" essential for the desired outcome in a two-way transfer
of information and ideas. As such, storytelling is very complex and communicatively demanding; it is so much more than just narrating the histories,
myths, and actualities of life. Beyond the narrating, one also uses distinct
forms of pitch, tone, specialized occasional expressions, amusing notes, for-
the-moment gestures, and all the repertoire of body language that could fix
or at least retain a substantial portion of the listener's attention.

In the case of using proverbs to communicate, advance, command, or at
times, deter specific behaviors, there is always a dash of poetry in creating
and verbalizing proverbs in the African context. Ruth Finnegan (1970) said
that in looking at the "structure of African proverbs, one of the first things
one notices is the poetic form in which many are expressed. This, allied to
their figurative mode of expression, serves to some degree to set them apart
from everyday speech" (p. 395). Needless to add that when proverbs are translated, the poetic notion may be lost or at least would not be as discernible as
in the original language. Even in textual societies, proverbs have the mini
distinction of being more oral, and thus closer to the cultural formations of
the group. Indeed, in these technologized times of early twenty-first century,
most old proverbs in the West seem to have been transformed into one line
or less than one line maxims that are, well, oral. One can actually see how
these have become nontextualized propositional or prohibitive notations,
with less binding power than proverbs in oral societies, but still effective in
their sociopsychological effects and outcomes. Interestingly, these maxims are
open-ended, referentially unbound, and safe from any copywriting entangle-
ments. As such, one may argue that they represent one aspect of the Western
communicative space that is fundamentally subjective, and therefore, more
humanizing, closer to life, and as they are freely exchanged even by strangers,
less calculating and more communal than everything else that characterizes
the nature of counting and accounting for realities in the so-called *adduunka
cusub* (*mondo moderno*).

Directly stating some examples[1] of proverbs from traditional oral societies in Africa should illustrate the widespread use of proverbs in valuing
or devaluing a situation, behavior, or expectation. Indeed, the following
proverbs, in different forms and structural constructions could be used in
other parts of the world, but what is good about all proverbs is that they
can always be contextually claimed and appropriated, which testifies to their
oral nature and that open sharing that orality usually connotes. On the
importance of leadership, the Hausa of northern Nigeria say, "a chief is like
a dust heap where everyone comes with his rubbish (complaint) and deposits

it." In the Somali context, there is *"nin xil qaaday eed qaad"* (a man who accepts responsibility, i.e., position of leadership, has accepted to be blamed for almost everything that goes wrong). On the importance of self-help, the Zulu (South Africa) stress, "no fly catches for another." The Somalis concur, *"nin walba tiisa haysata"* (every man [sic] is for himself). A more pointed saying on this same topic, which may be structurally more propositional than proverbial, is used by reer Kismayo (the people from the city of Kismayo in Somalia) which says *"Kismayo, Kistaa iyo Kaskaa"* (When in Kismayo, it is what you have (in money) and how wisely you use it). Here the embedded extra implication is how the people of Kismayo should be different from other Somalis who usually share what they have. The message warningly conveys, do not expect anyone to help you. In describing the highly disregarded nature of the so-called habitual liar, again the Zulu say, "he milks the cows heavy with calf." The Somalis have a simple description for that kind of person "bad *macaanshe*" (the one who transforms salty sea water of the sea into potable sweet water). In terms of not coming to terms with your reality vis-à-vis others, the Xhosa and the Zulu (both from South Africa) say, "there is no elephant burdened with its own trunk." The Somalis have a similar saying that also uses the elephant analogy: *"maroodigu takarta saaran ma'arkee, tan kaluu arkaa"* (the elephant does not see the insect on its body, it, indeed, sees the one on other elephants). Finally, as a way to relativize the real and difficult nature of life, and psychologically face it, the Nigerian saying, "no condition is permanent" has actually been globalized and used across many cultures and countries inside and outside Africa. To do something not entirely similar but related to the situation, the Somalis say: *"ká xun baa dhici kartá"* (we could find something worse in our immediate surroundings; i.e., do not exaggerate your problems, you could be in a much worse shape). Again, these proverbs effectively served, across time and space, as a quasi-codified jurisprudence where once the assumed precedents were stated, it was incumbent upon the people to refrain from pursuing any contrarian perspectives.

THE CENTRALITY OF POETRY IN ORAL TRADITIONS

The above-stated different methods of oral communication were usually functional in parallel terms with poetry, which was the most important form (in seriousness of the matter addressed) of expressing an idea, sanctioning a deed, or "authoritatively" predicting a multitude of social/other happenstances. However the poet conveys his or her message, the power of message was always superior to any other form of oral communication. Indeed, in all poetic transfers, whether the poem is transmitted verbally or via a written format, this form of communication still remains a powerful literature context that affects the lived situations of people. Still the saying that "poets are the unacknowledged legislators of the world" was more applicable,

I would instantly argue, to oral societies than traditions with a long history of written languages. Andrzejewski (1985) relates, for example, how poetry in oral societies influences all aspects of people's lives. This primacy of poetry, therefore, necessitated the memorization of poems that were composed for different reasons, events, and seasonal occasions. More importantly, poetic competitions and combative utilities of poetic eloquence were also common, where specialists memorized these poems, and regularly recited them to avoid mixing them with foreign material or even forgetting certain stanzas as time went by. This was also the case in conveying the poem or the narrative to the general or targeted audience. Andrzejewski (1985) noted how, like the modes of memorization, the channels of dissemination of works of oral literature can vary substantially.

Among the Somalis in East Africa, one the most distinguishing features of their poetry is that the lines must all contain one letter (j, d, g, etc.) which shows not only the unique brain power of the composers, but as well, the elegance with which the poem is transmitted and received. This unique nature of poetic composition might be specific to the Somalis, for I am not aware of anywhere else where this takes place. Indeed, it is not only brainpower, but also the richness of the language that provides an abundance of vocabulary that makes this possible. To understand and appreciate this qualitatively distinct nature of Somali poetry, which really affirms its linguistic superiority to anything produced in text-bound European quarters, we can look at the one below by Mohammed Abdulle Hassan, dubbed the father of Somali nationalism, who fought against the British between 1899 and 1921 (Samatar, 1982; Sheik-Abdi, 1993). In describing how he was traveling through difficulty terrain while trying to visit different homesteads to get support for his liberation struggle, Hassan (cited with translation in Andrzejewski, 1985), in his poem "the Jiinley" (i.e., the one that uses the letter "j" throughout; at least two words starting with "j" in each line), said:

Jafka hawdka guuraha, waxaan jar iska xooraayey
Wuxuu jeeni calaflow libaax, igu jibaadaayey
Raadkaan ku jiillaa, wuxuu daba jadeemaayey

(In the night journey through the dense forest, O how I stumbled down escarpments!
O how the heavy-footed lion roared at me!
O how he plodded after me, following the same footsteps as I!) (p. 351)

Looking closely at the Somali verses, it is not only that each lines contain words with the letter "j," it is also the case that the meaning of each line begins (rises, ascends) with a word that starts with a "j," and descends with another that again starts with a "j" without repeating the same two words in one line, stanza, or even in the whole poem. The emphasis that the poet places on the pain of his journey through thick forest where he is chased by lions, and where he is, therefore, in effect risking his life, is a call to arms. In other words, he was saying, "if this what I have to go through to transmit the message to you, my compatriots, then you should immediately realize the importance as well as the immediacy of the task, and rise up to liberate your country from the colonizing entity." Because of his poetic eloquence, which was mostly political and many times polemic, Hassan was so successful in recruiting thousands of men for his cause that he kept the British Army at bay for over twenty years.

Indeed, with his powerful poetic eloquence, and sometimes willful use of harsh terminology, Hassan kept on winning against the British until he was defeated by a new Western technology, air power. In 1920, aerial bombardment was first used in Africa when British planes flew from their bases in Aden, Yemen, and destroyed Hassan's military installations and control and command centers (Abdi, 1997). After that, Hassan fled to the interior of the rural country between Somalia and Ethiopia, but not without establishing, through poetry, the fact that his defeat was neither personal, nor factional, nor ideological. It was, he expounds in these lines, a national disaster that would affect future generations, and that the British only won because they corrupted people's social values and moral standards (please note that in this poem, the letter "d" is being used with the same recurring frequency as the "j" we saw above):

Dadow maqal dabuubtaan ku iri ama dan haw yeelan . . .
Nin ragay dardaaran u tahay, doqon ha' moogaado

Dawo laguma helo gaal haddaad, daawo dhigataane
Waa idin dagaaya kufriga, ad u dabcaysaane
Dirhamkuu idiin qubahayaad, dib u go'aysaane

Marka hore dabkuu idinka dhigin, dumar sidiisiiye
Marka xiga daabaqadda yuu, idin dareensiine
Marka xiga dalkuu idinku oran duunyo dhaafsadae

Marka xiga dushuu idinka raran, sida dameereede

(O, you people! Listen to these words of wisdom—or again you may simply
    ignore them, as you are wont to . . .
These are parting words of wisdom, and a warning to the wise—let fools
    forever remain in darkness!

There is no prosperity or peace that can come from entering into treaties and
    agreements with the infidel, for he cannot be trusted!
He is merely laying traps for you, while you let your guard down.
And the *Dirhams (money)* he dispenses to you now, will prove a poison in
    disguise!)

At first, he will disarm you, and render you defenseless like women and children,
Then, he will brand you, as though you were mere chattel

He will then press you to sell your lands to him for worthless trinkets;
And, finally, having dispossessed you, he will turn you into braying donkeys,
To bear his burdensome load.

                              Hassan, cited in Sheik-Abdi, 1993, p. 180

It should be useful to note how the poet's use of the term "infidel" to describe the colonial powers is designed to strategically locate his message and give it an otherwise untenable force in instigating the rural, relatively uneducated population who, in the poet's mind at least, may be more willing to fight provided that they sense a measure of belief sacrilege imposed by the foreigners. In terms of the point on "defenseless like women and children," in Somalia's nomadic warrior society, women and children did not bear arms, and were, ipso facto, defenseless in any context that was conflictual. Overall, though, and again from the ever present freedom of the expressive mode complemented by the absence of any imprisoning editorial complications, the poet appeals to his people using what was culturally treasured more than all else— liberty and land. Moreover, is the high probability that oral societies, because the verbal power is so essential for the formulation of thoughts, ideas, beliefs as well as the implementation of all agreed upon or situationally ordained tasks, possess a more refined and convincing expressive style that conveys all messages as clear, as strong, as specialized, and as urgent as they are intended. And as always, the presence of the inexplicably sophisticated mental process that "naturally" produces the letter "d" in the ascent as well as the descent of every line of the poem should be literally and analytically appreciated. As impressive as the unique poetic brilliance of this East African people, was the number of people in a given place or region, who were able to do it; it was not actually uncommon to find several people who could compose epic poems in one small village, complemented by few who were professional reciters.

It was, indeed, the peripatetic English man, Sir Richard Burton (1966[1856]) who called the Somalis a nation of bards (composers and reciters of poems). He was also very observant for a foreigner, when he described them as a fierce and turbulent race of republicans who are naturally anti-authority and predisposed to their individual kingdoms. Interestingly, the value of these observations has something in them that corresponds with the clash of oralities and textualities. Richard Burton wrote these when Somalis definitely knew these same qualities about themselves, but the perception of the latter were not institutionalized in a book. Currently, therefore, we could academically analyze the present Somali situation by referencing Burton, and concluding, rather hastily, that the reason they destroyed their state and cannot function, like other countries, in a national sovereignty-based global Westphalian system, is based on what the English man told us about them. Let me suggest that there may be more to the story than some simple historical unilinearities. The Westphalian system itself is text based and is driven by the values of geographically parametering written cartographies that have introduced to the Somalis and other oral societies the curse of fabricated nation-states (Davidson, 1993) that are culturally alienating, linguistically excluding, historically erasing, and politically disorganizing, all affirmed via mostly unsubstantiated written documents that tried to invent something called the Somali nation. No wonder, therefore, that today the conflict in Somalia contains something that represents a return to the primordial oral order, with all colonial and subsequent Somali government documents either lost or destroyed. So could this be a great refusal from the margins of the supposedly (if technically false) deoralized, and forcefully textualized? No, I actually will avoid any totalizing conclusions here, but perhaps we could delve, in another oeuvre, into this complex but important analysis of history, the word, the text, and their outcomes. Needless to add that other oral societies in Africa and elsewhere are hardly prospering. Only those segments that have been successfully conscripted into the culture of the text are materially endowed, mostly at the expense of the untextualized majority. So the clash of oralities and textualities is in full force, and will continue for the foreseeable future.

CONTEMPORARY TEXUALITIES FROM/INTO ORALITIES: SELECT ANALYSIS

Assuming that the current Westphalian system is presently the political order of the world, text-borne language also becomes the dominant mode of communication in especially those sectors of life that matter the most, which in this case are the political and economic spaces of the land. However, one can also say that by and large, the close-ended nature of written languages does affect the originality as well as the selective authenticity of the language that is being used in these crucial spaces of public life. In the political terrain, for example, the attainment of power is mostly based on the oratorial skills of the office seeker. So much so that in the so-called Western democracies, what is characterized as "one slip of the tongue" can ruin a very promising political career. As such, political language becomes the sine qua non of successfully winning political office and effectively holding it. Needless to add, holding political office is essential to the lives of people, for it is the space where public policy (or the management and distribution of public resources) is designed and implemented. Political office, therefore, determines what is important for contemporary societies, but especially for those who live in economically advanced countries where income and employment are two of the top three concerns (along with health care) that preoccupy people's daily existentialities. In the political speech, therefore, the common practice of relating to the subjective, multilocational, and complex lives of communities while reading from a prepared text is never as effective as having the primary oral skills that can reduce, even close the distance between speaker and audience. This validates the message as humanly and humanely constructed to address the tangible needs of people. Indeed, in the North

American context, for example, some politicians win office precisely because of the nontextual way they deliver their message. They are sometimes called "natural," that is, the one who comes as he/she is, speaks as she/he sincerely feels, and means what he/she says. Clearly, this supports the close range and on-the-spot constructed reality of oral communication. Indeed, the official political speech (commonly created by paid professional speech writers), is monotonous, repetitive, ideologically conscripting, and not responsive, either in structure or in tone, to the real needs of the public. Roger Anderson (1988) had a point where wrote:

> All political language draws on common experiences and tries to articulate new loci for identity and action. Overtly, political language is ambivalent and contradictory because *it is overused for so many divergent interests* . . . Who are the "we" and the "they" in speeches and in leaflets, posters and books? (p. 287, emphasis added)

With these apparent contradictions in the public communicative space, complemented by our pragmatic understanding that such real analysis will not realistically reenfranchise either the natural eloquence or the sociopolitical effectiveness of the oral tradition in places such as North America where the textual tradition has dominated for hundreds of years, the case may actually be different in formerly oral societies whose educational and official communication systems have been lately subsumed into the dominant text based traditions. Here, the psychocultural constructions of the communicative practice may be to a large extent still responsive to the oral tradition. In places like Somalia and other parts of Africa, it is not is uncommon to see politicians, talking to people without any written documents, but also bringing known orators who, in the conventional sense would be described as illiterate (read, nonschooled, and nonfunctional in connecting letters and numbers as reading/writing projects), but would be, in the way we are locating orality here, as historically, culturally, socially, and indeed politically highly literate in the contexts in which they reside. Politicians may also bring poets who recite newly minted poems about the candidate and his/her policy platforms, and it is here where the response is usually the strongest, and where we can critically see the intergenerational continuities of the oral tradition even after colonialism and its remaining programs have done to extinguish it and render it educationally and communicatively obsolete. The reality remains, that, however any system delegislates select means of communication or legislates others, orality as a cultural repertoire in people's lives cannot be texted out or inked out. Indeed, in certain spaces, when special interest oriented political expressions in written languages devour themselves, and the expressive sphere of linguistic operations cannot be maximized, one could call upon the elasticity of the points conveyed in the nonscholastic and nonrestrictive methods of oral communication that would be horizontally mobilized with prosaic impunity. Indeed, as Ashcroft et al. (2002) remind us, there is so much that we could learn from African oral literatures (and oral arts) in most spheres of linguistic expression, provided that we are willing to go beyond the epistemic misinterpretations of shallow anthropological exhortations. They write how the idea that

> African cultures had not, . . . developed writing beyond the earliest stages by the time of the colonial onslaught, should not serve to obscure the fact that African oral art had developed forms at least as highly wrought and varied as those of European cultures. Recognition of this led critics to urge that the study of these forms should be removed from the limiting anthropological discourse within which they were set and be recovered as a legitimate and distinctive enterprise for literary criticism. (p. 127)

ORALITY AND THE PROCESSES OF LINGUISTIC COLONIZATION

The perforce supremacy of text-borne colonial languages as the main methodologies of communication in colonial Africa and beyond, followed by the systematic relegation of historically and culturally rich oral languages that have defined, indeed sustained the lives of people (wa Thiongo, 1986; Abdi, 2008) to the status of epistemological invalidity, has created a situation where speaking English or French in Africa is equated with education, intellectual sophistication, and achievement. Here, the elite, who in relative terms mastered these languages (including the writer of this chapter), may not be willing to give back this culturally thin but materially thick status, for it accords them the means to earn a situationally more endowed living, but at the direct expense of the millions, who would have achieved much better in their life situations if the local languages they speak and use were as valued as the foreign tongues that continually affirm the deliberately concocted system of perceived ontological and learning inferiorities imposed on the general *vitae Africaines*. As Mazrui and Mazrui (1998) indicate, therefore, the organized inferioritization of mostly orally conveyed African languages have now an almost become permanent situation in which "one out of every five Africans on earth has a European language for a mother tongue" (p. 13). Clearly, the new English mother-tongued groups were not linguistically damned in a former life; definitely they had their own languages that were sociophysically created to respond and fulfill their relationship with their close and distant environments. And of course, the process of delinguicizing people should not be an easy one. It was indeed, achieved through the old, effective, and now almost proverbialized "carrot-and-stick" approach. One must actually give some credit to the colonizers for their brilliant understanding of the best way to assure the continuities of colonialism long after some form of "independence" was thrown to the way of the natives. The use of the term "independence" here may actually be insulting to any right-thinking person, relative to how the world is shaped today, where in actuality more entrenched systems of dependency have been achieved in the past fifty or so years; an issue that deserves some timely attention, but that will need a new chapter in a different book. As Thomas Macaulay (1999/1935) so cogently noted in the context of colonized India, the creation of an English speaking native elite who appreciate the ways of the English was to be the best way to continue the supremacy of Europeans over others. It was indeed, via this thoughtful analysis from the colonizers that the trajectory of emphasis in the project of colonialism was first and foremost, psycho-cultural-linguistic, followed by the more tangible political and economic dominations that most historians focus upon. It was also via the former that the relationship between the colonized and the colonized was collaboratively smoothed out and maintained for hundreds of years (Memmi, 1991).

Interestingly, the most effective ways of "delinguicizing" the natives, (with all the concomitant deculturing and socially dislocating outcomes they entailed) were the colonial systems of education that were not only deployed to advance the colonial project and its exploitative programs of learning, but also to literally rescind any viable spaces that could have been available for the use and development of mostly oral vernaculars that were used in pre-European hegemony Africa. With the revoking of African languages in spaces of learning and contexts of officialized life management, people and their primary means of defining and relating to their world were being torn asunder, establishing in the process, extensive regimes of mental colonization that are so much more difficult to overcome than any physical impact of the colonial project. In describing his personal experiences in his native colonial Kenya, Ngugi wa Thiongo (1993) noted:

> Our Language gave us a view of the world . . . Then I went to primary school and the bond was broken. The language of my education was no longer the language of my culture—it was a foreign language of domination, alienation and disenfranchisement. As a practice in colonial Kenya, anyone who was caught speaking the native language in the school vicinity was to be punished. (p. 11)

With the linguistic terrain and the oralities that carried it so marginalized in *Afrique noire*, one cannot ignore to ascertain the continuing identity and social development crises that expansively affected the lives of people. No nation, continent, or group can sustain the needed self-esteem and self-efficacy possibilities that are crucial for expansive, interconnected, and indigenously inspired human progress when the processes of deontologization, de-epistemologization, and delinguicization are so extensive and intensive. Indeed, the powerful Fanonian perspective that "a man [sic] who has a language consequently possesses the world expressed and implied by that language" (Fanon, 1968, p. 18) could not more appropriate here. And to respond to the possible inquiry of whether a language could belong to all those who may use it in their daily lives, I suggest that a language, be it English, French, Somali, Kiswahili, Burmese (Myanmar), or Vietnamese, has a history, it also has its value systems, moral sanctions, coded or quasi-coded interpersonal and intergroup strata, inherent emotional expressions, and special sentimental attachments, which all create a very special and unbreakable relationship between a language and the person who is native to that language. Indeed, the power of linguistic colonization is so strong in sub-Saharan Africa that even Ngugi wa Thiongo, one of the continent's most original writers and an ardent advocate for the Africanization of the educational terrain, had to retreat when he decided to write only in his mother tongue of Kikuyu. While he has not given up on the Kikuyu project completely, wa Thiongo's latest major writings are in English, and although practical issues surround these decisions, he did not leave the fight in reconstituting Africa's rich oralities and the life systems they carried for centuries. In his latest book, *Something torn and new: An African renaissance* (2009), he reexamines Africa's cultural fragmentation through colonialism, and especially emphasizes how what he calls "Europhonism" or the replacement of the African people's languages and identities with European ones, has led to the deformation of African memory. That memory was mostly carried in the continent's oral traditions and the prevalence of colonizing textualities will not help its revival. Indeed, these will continue marginalizing the space as well as the contents of that memory for the foreseeable future.

Pre-concluding Reflections

In terms of what could be immediately done to revamp the space of the oral, and thus rekindle Africa's memories, consciousness, and the confidence to achieve newly enriching viable spaces of social well-being, all the preceding arguments point to the minimizing of the textual power, the organized valuation of oral traditions, the reconstruction of educational spaces so they reflect the histories, the cultures as well and the actual needs of the African people. In reconstructing the educational terrain, especially, one needs to start with the recasting of philosophical platforms so as to raise the learning questions that horizontally correspond with the lived contexts of people. In addition, these learning systems should not continue diminishing the centrality of oral traditions to the lives of the African people. This does not mean that learners will abandon books and electronic sources that now contain the bulk of what they may be studying in conventional spaces of schooling. Nor does it mean that formal systems of education will be en masse replaced by something new that introduces a paradigmatic shift in the relationship between students and contemporary institutions of education. As I have discussed previously (Abdi, 2002), the Africanization of knowledge intends to achieve, especially in these times of the *post-facto* reality, novel but fundamentally inclusive prospects of learning where the best from European style of education would be intermeshed with African theories of knowledge and ways of knowing. Cleary, African theories of knowledge and ways of knowing carry the memories of orality, thus the reconstituting of stories, proverbs, and poetry has to be reenfranchised as important to education and social development. Perhaps more important is the overdue redemption of African languages as important disciplines that should be studied and selectively used as media of instruction in as many learning platforms as practically possible. Knowledge is both a collective human project and a social construction that is selectively reflective of people's lives in given tempo-spatial possibilities and intersections. As such, the continuing epistemic marginalization of so many in Africa and elsewhere cannot and should not continue.

CONCLUSION

In this chapter, I have briefly introduced the oral ways of sub-Saharan Africa and discussed the perforce triumph of textualities via the colonization of the general communicative space. In analyzing these, I spoke about the richness of oralities, especially in the way they represented the reality of people's lived contexts. Indeed, the closeness of oral expressions to social situations and relationships is something that even Western researchers have accepted, and as such the strength of the subjective presentations and tangibilities of orality should not beyond doubt. Despite all of this, colonialism, in demeaning and erasing indigenous ways of learning and knowing, was adamant on fully establishing the supremacist perspective of the written word, thus minimizing the narrative as well as the poetic elegance of oral traditions developed over millennia. As these processes of cultural and espitemological alienations expanded in the lives of the colonized, it is clear that the level of ontological and practical livelihood damage was extensive to the extent where we could argue that the imposition of textualities on people with different traditions was destructive and has almost permanentized problematic life spaces in identity, self esteem, and overall social well-being. Especially in the linguistic terrain, the triumph of textualities was responsible for the stunted development of many African languages as speakers of these languages were systematically delinguicized and burdened, via externally imposed economic relationships, educational oppressions, and outright punishment for speaking their mother tongue, with existentially depriving sets of English, French, Portuguese, and other tongues of the European metropolis. While this is what has happened, it does not mean that this is how things have to be. In the new possibilities to recast Africa's cultural, linguistic, and overall development terrains, the best option may be to aim for multicentric ways of learning where mixing the best from both worlds could serve better intentions and effective learning outcomes. Indeed, the relivening of Africa's rich oral traditions would contain so much social, environmental, and technical knowledges and know how, that by collectively investing in their multilocational enfranchisements, the whole world would do much better in all aspects of global ecological well being. As other epistemic representations, languages, and methodologies of perception and explanation, Africa's thought systems, styles of knowing, and oral ways of creating and using select knowledge plateaus, can be and should be reconstructed as important educational and social development platforms that benefit all.

REFERENCES

Abdi, A. A. (1997). The rise and fall of Somali nationalism: From traditional society to fragile nationhood to post-state fiefdoms. *Horn of Africa Journal, 25*(1–4), 34–80.

———. (2002). Postcolonial education in South Africa: problems and prospects for multicultural development. *Journal of Postcolonial Education, 1*(1), 9–26.

———. (2008). Europe and African thought systems and philosophies of education: "Re-culturing" the trans-temporal discourses. *Cultural Studies, 22*(2), 309–327.

Achebe, C. (1996 [1958]). *Things fall apart*. Oxford: Heinemann.

Anderson, R. (1988). *The power and the word: Language, power and change*. Toronto: Paladin Grafton Books.

Andrzejewski, W. B. (1985). Somali literature. In W. B. Andrzejewski, S. Pilaszewicz, & W. Tyloch (eds.), *Literatures in African languages: Theoretical issues and sample surveys* (pp. 337–407). New York: Cambridge University Press.

Ashcroft, B., G. Griffiths, & H. Tiffin (2002). *The empire writes back: Theory and practice in post-colonial literatures*. New York: Routledge.

Bourdieu, P. (1991). *Language and symbolic power*. Cambridge, MA: Harvard University Press.

Brown, G. (1995). *Speakers, listeners and communication: Explorations in discourse analysis*. New York: Cambridge University Press.

Burton, R. (1966 [1856]). *First footsteps in East Africa*. New York: Praeger.

Fadiman, G. (1994). *When we began, there were witchmen: An oral history from Mount Kenya*. Berkeley, CA: University of California Press.

Fanon, F. (1968). *The wretched of the earth*. New York: Grove Press.

Finnegan, R. (1970). *Oral literature in Africa*. Oxford: Clarendon Press.

Habermas, J. (1998). *On the pragmatics of communication*. Cambridge, MA: MIT Press.

Hegel, G. W. F. (1965). *La raison dans l'histoire*. Paris: UGE.

JanMohammed, A. (1983). *Manichean aesthetics: The politics of literature in colonial Africa*. Amherst, MA: University of Massachusetts Press.

Kaplan, D. (2003). *Ricoeur's critical theory*. Albany, NY: SUNY Press.

Krog, A. (1998). *Country of my skull: Guilt, sorrow, and the limits of forgiveness in the new South Africa*. New York: Times Books, New York.

Macaulay, T. (1999 [1935]). Minute on Indian education. In Ashcroft et al. (eds.), *The post-colonial studies reader* (pp. 428–430). New York: Routledge.

Mazrui, A., & A. Mazrui (1998). *The power of Babel: Language and governance in the African experience*. Chicago: University of Chicago Press.

Memmi, A. (1991). *The colonizer and the colonized*. Boston: Beacon Press.

Nyerere, J. (1968). *Freedom and socialism: A selection from writing and speeches, 1965–67*. London: Oxford University Press.

Ong, W. (1985). *Orality and literacy: The technologizing of the word*. New York: Methuen Press.

Pellowski, A. (1977). *The world of storytelling*. New York: R.R. Bowker Publishing Company.

Rodney, W. (1982). *How Europe underdeveloped Africa*. Washington, DC: Howard University Press.

Said, E. (1993). *Culture and imperialism*. New York: Alfred A. Knopf.

Samatar, S. (1982). *Oral poetry and Somali nationalism: The case of Sayid Mahammad Abdille Hasan*. Cambridge, UK: Cambridge University Press.

Sheik-Abdi, A. (1993). *Divine madness: Mohammed Abdulle Hassan (1856–1920)*. London: Zed Books.

Swanson, D. (2007). Ubuntu: and African contribution to (re)search with a "humble togetherness." *Journal of Contemporary Issues in Education, 2*(2), 53–67.

wa Thiong'o, N. (1986). *Decolonising the mind: The politics of language in African literature*. London: James Curry.

———. (1993). *Moving the centre: The struggle for cultural freedoms*. London: James Curry.

———. (1998). *Penpoints, gunpoints and dreams: Towards a critical theory of the arts and the state in Africa*. Oxford: Clarendon Press.

———. (2009). *Something torn and new: An African renaissance*. New York: Basic Civitas Books.

Autonomy and Video Mediation: Dalitbahujan Women's Utopian Knowledge Production

Sourayan Mookerjea

Introduction: Subaltern Knowledge-Power

This chapter examines the (re-)production of subaltern or "indigenous knowledge" through video-based media of a group of Dalit women farmers from Andhra Pradesh, India, who belong to a media cooperative they have named the Community Media Trust (CMT). The women of the CMT are a vital and active body in the network of *Sangham* (agricultural cooperatives) to which they belong, and their media is politically engaged: Agricultural, environmental issues, particularly pertaining to biodiversity, seed sovereignty, and gender issues, like subaltern women's empowerment and dignity, are constant themes of their work. They have made point-of-view documentaries, grassroots investigative journalism, promotional videos showcasing development projects, as well as participatory research videos in collaboration with the Deccan Development Society (DDS). Drawing on all these genres but going beyond them are a group of videos, which are crucial to their collective political project of managing their cooperatives and producing, reclaiming, and conserving subaltern "indigenous knowledge." My focus in this chapter will be on this particular aspect of their media practice. The women of the Sangham seek to achieve autonomy over food production, access to seeds and other natural resources, markets and, through all of these, autonomy in their livelihoods and for their community's future. Their endeavors to create an autonomous mode of knowledge production, as we shall see, is crucial to these struggles and the CMT's quest for autonomous media, as we shall also see, is crucial to all their struggles for autonomy.

This chapter is organized in three parts. I first provide some historical background that locates the women of the CMT and their primary audience, the marginalized farmers of the region, in relation to the persistence of class, caste, and gender inequalities of postcolonial India and with respect to their contemporary struggles against exploitation and domination at the hands of globalized agribusiness and the neoliberalized Indian state. The second section describes the institutional background of the CMT and the DDS as well as the main features of their quest for autonomous media. The final section focuses on the mediation of subaltern knowledge. Here, I examine the specific political-aesthetic features of the CMT's practice of autonomous video that enables subaltern knowledge to be conserved through its production and politicization. Critical scholarly studies of subaltern social movement media practices can provide crucial insight into the mediation processes through which subaltern social movement education and knowledge production unfolds.

Before proceeding down this path, however, a preliminary terminological note is in order. While the term "indigenous knowledge" has global currency, especially in relation to struggles against biopiracy, its use in the subcontinental context presents several widely noted difficulties. On the one hand, the adjective "indigenous" is often used to describe anything assumed to be either local or non-European in provenance, in accordance with one of the received definitions of the word. On the other hand, the term "indigenous" is used to translate the term *adivasi* and is then used synonymously with the Indian English term "tribal." But this usage is also problematic for several reasons. As Hardiman (1987) notes, the neo-Sanskritic term *adivasi* is itself an invention of missionary intervention in the Chotanagpur district of Bihar in the 1930s. This missionary reclassification of polity and society was moreover overdetermined by the racializing nomenclature invented and imposed by the colonial state in response to the Santhal Insurrection of 1855–1856, the 1857 Revolt, and the Sardare Movement of 1859–1865, and through subsequent censuses and ethnological surveys (Bates, 1995; Karlsson, 2003; Pinney, 1990; Skaria, 1997; Xaxa, 1999). The colonial race classification hierarchy, as a key administrative tool of governmental rationality, then becomes a historical mediation of all subcontinental modes of social reproduction through and against which identities come to be formed and reconstructed. Moreover, as Beteille (1998, 2006) and Guha (1999) argue, these colonial constructions take place on the cultural common ground of long-standing historical communication, migration, and interaction between subaltern cultivators and forest dwellers living on the margins of Hindu and Muslim kingdoms of the precolonial era. Moreover, both *dalit* civil rights movements and Hindutva neo-Fascist nationalist movements have self-identified as movements of indigenous people (Baviskar, 2006). Consequently, the definition of ethno-cultural, caste, religious, and production mode differences of identity are as problematic as they are multitudinous on the subcontinent. Indeed, such differences—even of modes of production—are sites of political struggle, not only of cultural distinction (Shah, 2007). Furthermore, in recent decades, an anticommunalist and anticapitalist *dalitbahujan* (the term names the dalit or subaltern majority of people in India) politics has emerged that seeks to forge counterhegemonic alliances between different subaltern locations. Also, the question of whether even *adivasi* struggles on the subcontinent can be best carried out by articulating the "indigenous slot" (Karlsson, 2003, p. 404) at international fora remains open. For these reasons, I will not use the term "indigenous knowledge" in this chapter, preferring the term "subaltern knowledge" instead; and my study of the CMT's video practice will endeavor to show that subaltern knowledge produced in the course of a social movement struggle is to be even more precisely understood as *Utopian subaltern knowledge* in the strongly positive sense of the Utopian global social justice politics of the current conjuncture (Jameson, 2005).

MEDAK AND THE MULTINATIONALS

The politics of autonomy pursued by the women of the Sangham responds to changes in the modes of domination of small farmers in postcolonial India and cannot be understood without at least a cursory overview of these conjunctural developments. As Green Revolution ideologies and practices diffused beyond India's northwestern belt, farmers increasingly became dependent upon a national network of agricultural research institutes, public authorities, and private companies as well as the technocratic expertise embodied by this apparatus of agricultural management. This dependency has two crucial aspects to it that are both fronts of struggle in the Sangham women's quest for autonomy. There is, first, a dependency rooted in access to crucial farm inputs, especially seeds and fertilizer, as seeds bred by research institutes and private companies displace those which used to be collected and saved by farmers themselves. Seeds and pesticides would now also have to be purchased along with industrially produced petrochemical based fertilizers. Second, these new components of agriculture now require monopolized instructions on their use, which renders obsolete whatever agroecological knowledge farmers have traditionally taught down generations. Over the 1980s, national and multinational agribusiness corporations gained greater access to markets created by this managerial complex (Patel & Müller, 2004). With official liberalization of the Indian economy in 1991, the multinationals have gained even more influence over agricultural policy.

This dependency has been exacerbated of course by the abysmal failure of the Indian welfare state to deliver social services to the countryside. Not only are large numbers of Indian farmers left nonliterate, but their non-literacy exacerbates their dependency on the managerial complex of the agribusiness multinationals to the same degree as it secures the managerial complex's monopoly of technical information over the farmers. Moreover, a further legacy of the national development era in Indian agriculture is the "modernization" of caste and patriarchal ideologies turned virulent by the uncertainties and anxieties provoked by the faceless chain of dependencies of capitalist agriculture.

The Dalit farmers of Medak, especially women, are thus the most marginal of the marginalized. Caste discrimination during the postindependence land reforms left them with the least arable land. Patriarchal inheritance norms result in many Dalit women becoming landless and most enter farm wage labor, as sexist and caste assumptions underlying the division of labor, along with illiteracy bar these women from many other better protected occupations. The steady retreat of the state from rural development and privatization in the agricultural sector, however, has furthermore decreased the availability of rural employment. Men and women continue to leave the countryside, making for a ready supply of low wage, politically vulnerable labor in the cities. Women in particular are absorbed in the informal sector, especially in domestic service as well as the poorly regulated construction industry where many women are employed building India's high tech IT parks (Patel & Müller, 2004). The contradictions and conflicts between rural and urban India (and ultimately, the urban world economy) thus continue to intensify. For the rural crisis is of considerable benefit to the more privileged corners of the world economy as the lowest wages in any economy keeps the cost of all other wages down and the relayed savings is thereby available to trickle up toward any effective monopolies as privately appropriated profit (Wallerstein, 2004). But whether the Dalit women remain in agriculture or move to the cities, there are two other aspects of the postnational development era of rural crisis that bear significantly on both the women's readiness to accept wages of poverty as well as the Sangham women's strategy of autonomy.

The first of these is the World Bank's sustained attack on the Indian Public Distribution System (PDS) for Foodgrains. Citing problems of bureaucratic ineptitude and corruption, but also for ideological and strategic reasons, the World Bank leaned heavily on the Government of India to scale back the PDS. In 1992, the government caved to these pressures, and in the name of efficient targeting of the poor, tightened access to the program and reduced its scope. Most commentators and critics of the Indian PDS agree the program was in need of reform. But the World Bank's prescriptions have only increased the endemic hunger of the poor and worsened the incidence of malnutrition, especially of women and children (Patel & Müller, 2004). The Sangham women have consequently organized their own Alternative Public Distribution System and the CMT has made videos that explain how this program works and advertise its availability.

The Sangham women also confront the determination of large agribusiness, led by Monsanto, to impose a second green revolution—sometimes called the Gene Revolution—on Indian farmers. American biotechnology companies have battled long and hard to get markets created for their products in India and obtain the regulatory changes that would allow this. One major victory in this campaign was the 2002 National Seed Policy followed by the Seed Bill of 2004. The former completed a process begun in the New Policy of 1988 and fully opened the door to the agribusiness multinationals to market their seeds in India by freeing imports and exports of all seeds. The Seed Bill of 2004 went further by making seed registration mandatory for farmers who saved, exchanged, or sold their seeds for agricultural purposes. In effect, the bill made the traditional practice among small farmers of saving and exchanging seeds illegal (Kumbamu, 2006). The bill thus not only sought to secure a monopoly space for agribusiness seed producers but also posed the danger that seed registration would facilitate biopiracy wherein biotechnology companies appropriate indigenous agromedical knowledge and gain intellectual property protection for such knowledge in the United States. But it also seemed certain that without seed registration, biopiracy would take place anyway. These regulatory changes consequently struck at what little autonomy the small farmer retained after the reorganization of agriculture by the first Green Revolution. Again the Sangham women have responded by developing their own agrobiodiversity register and some of most important media work done by the CMT are the videos they have made on this issue.

DALIT WOMEN'S SANGHAM AND THE CMT

In the face of such conjunctural transformations, Dalit women in Medak began to organize themselves in the mid-1980s into affinity groups or voluntary associations for organizing microcredit funds in order to rent fallows. In 1983, a group of social scientists and community development activists founded the DDS and began working in Zaheerabad, Medak on converting a development project abandoned by a private company into environmentally sustainable rural employment by bringing stony and degraded fallows into cultivation. The DDS joined with the women and arranged a state start-up grant for their initiative. The Sangham network of agricultural cooperatives to which the women of the CMT belong have their roots in these "chit fund" groups. The women's Sangham are now active in about seventy-five villages and over five thousand women belong to them. In the intervening twenty-five years, the Sangham have brought under organic and biodiverse cultivation over ten thousand acres of degraded land and produce over six million kilograms of local millets, sorghum, and pulses annually. Over the years, the Sangham women have also assumed responsibility for more and more aspects of the everyday operations of the DDS and now form the major part of its think tank and core management team.

Along with daycares and a school, the Sangham network also manages several forest commons of over a thousand acres, which they have regenerated near their villages and in thirty villages they maintain medicinal commons where over sixty different species of medicinal plants are conserved. Another key autonomy struggle against biopiracy and seed monopoly has led to their invention of the Agro-biodiversity Register and the establishment of Community Gene Funds in sixty villages where more than eighty species of cultivars have been retrieved from extinction and conserved as collective public property. Designed by themselves for themselves and other nonliterate farmers, the Agro-biodiversity Register has been adopted into the Indian National Biodiversity Strategy and Action Plan.

The DDS-Sangham link thus constitutes a political body that has the configuration of a network of networks. On the one hand, the DDS belongs to several environmental and national, regional and international antiglobalization coalitions and solidarity networks such as the Organic Farming Association of India, the Southern Alliance Against Genetic Engineering, South Asian Network For Food, Ecology and Culture (SANFEC), South Against Genetic Engineering (SAGE), as well as Biodiversity Action for Sustainable Agriculture (BASA-Asia). In this, we find traces of the peculiar twist the Indian national liberation movement gave to Fanon's famous primal scene in which the student from the city escapes to the countryside and finds shelter and political enlightenment in the hospitality of the subaltern and a revolutionary decolonization process thus comes into being: The agronomists, biologists, social scientists, journalists, environmentalists, policymakers, and activists from urban India and beyond who attend the DDS's consultations, hearings, workshops, and public fora all walk paths historically laid down by both the Gandhian and Communist mass movements' tireless organizing work among the rural poor. Faced with the neoliberal offensive as well as assertive Dalit self-organization from the 1980s on, both movements have been forced to rethink their politics from first premises and the DDS is one conjunctural experiment in this critical process. Through this organizing work, the Sangham has succeeded in opening a site of struggle within the Indian state's managerial complex for agriculture. The risk of cooptation involved then makes the Sangham network all that more crucial to the women's struggles. The Sangham network, on the other hand, has not only raised the status of the women who belong to it and enabled them to assume many kinds of leadership roles in their villages but their pedagogic and recruitment efforts—especially through their organization of an annual biodiversity *jathara*—have extended these women's influence throughout a wider class formation. The Sangham network has also become more *bahujan* in character by forging solidarities with a broader class of poor women including Muslims, Gollas (a cattle breeder caste), Tenugus (marginal peasants and fisher-folk), Mangalis (a caste of barbers), and Sakalis (a caste of washers). As we shall see, "the network-of-networks" (Hardt & Negri, 2000,

p. 295) character of this social and political body forms a crucial context in which the CMT was formed and now works.

In order to consolidate and further their pursuit of autonomies, the Sangham women, in particular their elders, resolved to create their own autonomous media. Ten women completed a year long training program organized by the DDS in a series of media workshops led by P. V. Satheesh and another professional media producer in 1998–1999. The DDS also built and maintains a production studio that the women use. The CMT was formalized on International Rural Women's Day (October 15) in 2001 when a board of eleven trustees, including eight Sangham women, assumed management of the media cooperative.

The relationship between the network of Sangham, the DDS and the CMT is defined by the struggle for small farmer autonomy. A key strategy in this struggle for autonomy is the elaboration of a kind of agriculture that is petroleum free, that conserves biodiversity, and that is adapted to water scarcity. To this end the DDS runs the Krishi Vigyana Kendra, an agricultural research institute where the women of the Sangham, development workers, environmental scientists, and agronomists from around the work collaborate on research. An important dimension of this collaboration is the documentation, adaptation, and dissemination of what the DDS calls "traditional knowledge systems" and "indigenous technology."

THE POLITICAL-AESTHETICS OF THE SANGHAM SHOT

In order to gain a deeper understanding of the singular features of the CMT's autonomous media and its importance for the ongoing viability of the cooperatives, let us begin with the women's own videographical manifesto as they present it in their video *The Sangham Shot*. This video begins with an aesthetic political statement. The videos these women make, we learn, are organized around the formal principle of an eye-level shot which they call the "sangham shot," since in the Sangham, the women tell us "we are all equals." They distinguish this shot from two others: the "patel shot" (or landlord shot) where the camera looks from on high down at the women working the soil and what they call the "slave shot," which is the reverse shot taken from below looking up. I will return to consider the significance of this distinction later as it goes beyond the fact that at public meetings of the Sangham or the Gram Sabha, people sit on the ground as they do in the course of performing many agricultural and domestic tasks. However, several of their videos make it clear that the sangham shot is indeed an aesthetic and political principle informing their work, equally so for those videos that seem to have been made primarily for farmers in their district (such as *Making of an Agricultural Biodiversity Register*, *Our Watersheds, Our Priorities*, and *Onwards to Food Sovereignty*) and for those which seem to be made primarily for their solidarity network (such as *Water, Life and Livelihood, Sustaining Local Food Systems*, and *In the Lap of Pacha Mama*).[1]

In both kinds of videos, typical "talking head" explanations are replaced by the sangham shot's eye-level, mid range face-to-face frame of someone making a statement or declaration about a particular problem or issue. The sangham shot does not make use of any conventional "Gutenberg" props to legitimize literate expertise (bookcases, desks, podium, microphone, computer screen, etc.). Nor does it frame an ideal type (of the nonliterate farmer) or a general condition (e.g., of water scarcity). Rather, it always gives us a specific person from a specific village speaking about this field, that well, those crops, or these problems. Indeed, Chinna Narsamma of the CMT emphasizes just this point in one of her statements in *The Sangham Shot*. This is not to say that their videos abjure explanations and generalizations altogether; they do not. They are used when necessary (as in the documentaries on water issues). But most often such elements are syntactically subordinated to an acoustic-image composition of a direct, face-to-face, immanent address meant to catch and hold the viewer's look.

In the important video *Making of an Agricultural Biodiversity Register*, for example, this syntax plays a particularly crucial role as the video documents extensive community involvement in making up an agro-biodiversity register. Both individual involvement and community cooperation are foregrounded this way, as are women's leadership and expertise in the process. In its efforts to advocate the making of such biodiversity registers as a seed sovereignty and autonomy strategy, the video explains how the village of Khasimpur made its biodiversity register and presents this as an example for other subaltern farming communities in the district and beyond to follow. For the Sangham's solidarity network of outsiders, this explanation takes the form of English subtitles outlining the steps. But this is the sketchiest of explanations, as it becomes clear from another form of writing the farmers compose on the grounds of the village assembly out of flour, vegetable colors and dyes, seeds, plants, and terracotta figures that extensive knowledge of dryland agriculture is presupposed in this account. This, however, allows the video to drive home another key point it sets out to make. Not only are the assembled farmers experts in biodiverse agriculture, but this expertise belongs to the community. As the editing cuts back and forth between the colorful register on the ground and the public deliberation between farmers around it, the sangham shot's face-to-face immanence serves to show how this expertise is cooperatively sustained. The video concludes with the transcription of the register from the ground to a book and the Gram Sabha's (village assembly) certification of the veracity of the book's contents.

As a viewer, it becomes very difficult to extricate oneself from the appeal of face-to-face immanence, to turn one's back, so to speak, and file away what you have just seen and heard as ready-to-use information. Rather, the sangham shot demands concentration and response. We will need to come back to this aspect of the CMT aesthetic below but for the moment we can note the obverse feature of the sangham shot which partly accounts for the kind of demand these videos make and the pleasure they give.

For face-to-face immanence can be inverted by the sangham shot to illuminate a singular kind of movement—the movements of an embodied relation—as this makes up petroleum-free agriculture. The video *On Women and Genetic Diversity* provides a paradigmatic demonstration of how the CMT inverts the sangham shot into an image of such movement. The video is composed primarily of various movements of biodiverse agricultural production as it is performed by women especially. These include shots of women saving and storing seeds, varieties of seeds, the use of a drill to sow seeds, crops at various stages of growth, rain, women collecting seeds. These shots build up to profile two community elders in particular, Gangwar Anjamma and Gangwar Manemma, women who are deeply respected by the CMT (a third farmer also shows us the value of multistoried planting). The series of movements in production tracked by this video are "completed" into some kind of unified or continuous movement making up the video's beginning, middle, and end, however, by the two women's direct statements to the camera. Gangwar Anjamma tells us that she never purchases seeds but that she saves them using neem leaves and ash, nor does she ever use petrochemically derived fertilizers. Gangwar Manemma, standing among her crops, tells us that she grows twelve varieties of crops and that even if some of them fail, she is always able to feed her family.

All the videos I am discussing here document and analyze various movements of petroleum-free agricultural production. *Our Watersheds, Our Priorities*, another video made primarily for other farmers, documents the tasks of destoning land and building rockfill dams and bunds in a watershed self-management project. These embodied movements of people in their environment are connected to the face-to-face immanence of a watershed community meeting held in Edakulapally, Medak District (March 17, 2004) where men learned from women farmers about vermicomposting for supplies of manure and where the women resolved to make a video about their watershed management plan to demonstrate that this is their own effort, not a government project. This video concludes with the farmers pledging to grow food crops instead of cash crops and with a videotaped certification of the consensus reached by the meeting. Two videos on the Sangham women's appropriation of technology, *A Machine for Millet* and *Biofertilizers: A Technology Brought Home*, provide in-depth analyses of movements of production. The first of these explores the problems posed by the time-consuming and arduous nature of grinding millet into flour and the Sangham women's consequent collaboration with engineers on designing a millet mill that they run as a cooperative serving the wider region. The biofertilizers video details various indigenous soil fertility practices involving green manure, tank silt, cultivating legumes, farmyard manure, and vermicomposting (which Sangham women use themselves, selling the surplus compost through a cooperative).

I should underscore the point that the videos' analyses of such agricultural techniques lie in the CMT's composition of movements out of both the camera shots the women are able to get as well as through their editing. (To assert just this, all of their videos include shots of themselves videotaping). In both, camera-work and editing, their agricultural practitioner's expertise becomes indispensible. The intelligibility of the videos' analyses rests with the composed visibility of the movements of production.

But perhaps the most compelling achievement of this movement composition to date is to be seen in their collaborative video, *Using Diversity*, where the women of the CMT teamed up with women videographer-farmers from Nepal, Bangladesh, Maharashtra, and Sri Lanka. My prose here can only clumsily suggest one's rapt, embodied relation to the movement of the video-image in this exemplary sequence of shots: building bunds, bullock plowing, performing *puja*, transplanting rice, videotaping, all of which seem merely quaint and stereotypical when stripped of movement's timing. This is not merely because written prose can only inadequately translate video images. Or rather prose fails here because these videos make visible a sense of timing rooted in oral culture. They bring to the foreground of our perceptual gestalt what Harold Innis (1951) called orality's time bias of communication.

Innis' communication theory proposes that insofar as every media of communication interferes with our being in time and space and recomposes those relations in some way, every media of communication then involves a bias toward either space or time, insofar as we are unable to maintain a balance between either qualities. But against technological determinist readings of this thesis, such as McLuhan's, I want to suggest that Innis' understanding of mediation is rather an immanent one. That is to say, for Innis, every media presupposes a determinate set of social relations of communication involving some specialization of expertise and capability (even to the point of there resulting in some monopoly). But then every media is thereby involved in the production of new spatial and temporal relationships in which yet new social practices flourish.

For example, in formulating his critique of the Cold War world order and its "military industrial complex" as it emerged in the last decade of his life, Innis studied several historical processes unfolding over the nineteenth century, particularly in the United States, and drew together their combined significance. The rise of mass publishing and newspaper-chain monopolies not only left a nation-building footprint in their need for a sufficiently large market to realize profits. But the very scale of their competitive operations also drove them along the same path as every other kind of industrial capital toward monopoly formation. Meanwhile, the industrial division of labor in formation was not only the arena and the result of class struggle and capitalist competition but for this reason also abetted the monopoly character of capitalism; so much so that the United States invented a specific institution to ride these trends—the corporation. Designed to politically organize and defend monopolies of expertise, they would of course play a decisive role in enabling the United States to usurp the British Empire (Arrighi, 1994). Innis was especially concerned about several further attendant features of these processes in which the publishing industry was particularly implicated. He noted, first, the press' leading role in warmongering, which was decisive in the American state realizing its imperialist character. Second, Innis pointed to the press' hand in restructuring the office of the presidency and in undermining the democratic promise of the Constitution (Innis, 1952). Last, mass publishing together with the corporate institutional form of the division of labor resulted in a very particular form of literacy reduced to and organized around information. All of these historical processes together constituted what he termed the space bias of industrial communication:

> Intellectual man of the nineteenth century was the first to estimate absolute nullity in time. The present—real, insistent, complex, and treated as an independent system, the foreshadowing of practical prevision in the field of human action—has penetrated the most vulnerable areas of public policy. War has become a result and a cause of the limitations placed on the forethinker. Power and its assistant force . . . have become more serious since "the mental processes activated in the pursuit and consolidating power are essentially short range". (Innis, 1952, p. 41)

Not only does Innis' critique hold for the communication system designed by the multinational corporations and their states in our own era, but Innis' assessment of its antidemocratic character and its unsustainable nullification of time accords very cogently with the Sangham women's own experience of monopolistic agribusiness and postliberalization media in India.

But the point I want to make here is that video's ability to travel speedily and conveniently into the solidarity laceworks and its communicative effectivity in the world of literacy is its link to information. The (post-)industrial, international social division of labor calls into being both class struggle and information as its characteristic features and video remains tied to both, as so many community media activists around the world have demonstrated. But along with this spatial property of video mediation, the sangham shot's aestheticization of movement also retrieves and refunctions orality's bias of time. The women's camera "sees the world" just as the primary producers of this kind of agriculture see and understand the relationships between each activity (plowing, sowing, weeding, harvesting, cooking, nurturing); we also see how each *becomes* the other. Such analyses of movement in all these videos are nothing less than Taylorist time-and-motion studies carried out on the basis of some other principle than the efficiency of domination in the production process.

There are two further aspects of the movements of production (and their reliance upon the time bias of communication) these videos show us that we need to note here. The first is the impression these videos create—complementing the videos' insistent polemical assertions—that the agricultural practices we are seeing (bullock plowing, hand winnowing, hand weeding, etc.) are definitely neither archaic agricultural practices nor soon to be obsolete movements of production. Rather these movements compose some kind of Utopian present-future liberated not only from petrochemical dependence nor from those economic and political dependencies petroleum-based agriculture here presupposes, but from all the dependencies which have been accumulating in the world ever since the Industrial Revolution fatefully introduced a host of spatial contradictions and conflicts between town and country. This is why the movements that make possible petroleum-free agriculture must also be sung whenever they are shown, since this present-future is also that of an oral culture. The videos' accounts of the various dimensions of autonomy that the women of the Sangham are pursuing achieve their compelling force in this way.

Second, the aesthetic-cognitive experience of this kind of communication (beyond whatever information may be conveyed, whatever representations may be made with respect to its space bias) is itself made up out of a repetition of several movements that we need to note here. Our aesthetic experience as participants of this communication involves the movements the women's bodies are capable of as practitioners of biodiverse, petrochemical-free agriculture, the movements they are then able to compose with video images, and the 'tidal movements' of affect and thought these are able to produce in the bodies of their audiences. It is in this articulation that the time bias of communication asserts its powers. This is what the aesthetic-politics of the sangham shot is capable of achieving. This is why it is distinguished from the shots the CMT call, significantly, slave and patel shots. The sangham shot is able to bring this time bias of communication to the foreground of the video image movement. In doing so, these videos demand not our recognition but rather our concentration and our response. Indeed, this is what the production and conservation of subaltern knowledge entails: the *feminist* constitution of a political body in movement; and this is why it is a species of Utopian knowledge.

UTOPIAN SOLIDARITIES

This study has tried to show that the video practice of the CMT is both crucial to the organizational work of the Sangham's self-management and to the production of agriecological and political knowledge to this end. The CMT's video practice is but one mediation process on which subaltern knowledge production here depends, but it does vividly demonstrate that such knowledge is Utopian in the precise sense that its social and political actuality and its aesthetic effectivity escapes and so deconstructs the ideological binary oppositions between traditional knowledge and scientific knowledge, or indigenous knowledge and modern knowledge, or Western knowledge and so on, that have structured so much of our discussions of subaltern politics and agency. But in calling the CMT's video aesthetic Utopian, I am also drawing upon and bringing together not only Harold Innis' theory of the bias of communication but also Fredric Jameson's (2005) demonstrations of the Utopian figuration at work in any collective political unconscious. Indeed, we could easily miss the whole Utopian process in play in this media "unless we grasp its critical negativity as a conceptual instrument designed, not to produce some full representation, but rather to discredit and demystify the claims to full representation of its opposite number" (Jameson, 2005, p. 175). Its moment of truth is therefore "not a substantive one, not some conceptual nugget we can extract and store away, with a view towards using it as a building block of some future system" (p. 175) since, unlike the agribusiness model, it does not teach a monological, nonholistic, present-minded method. Rather its truth resides not only in the singular survival strategy it enables but also in its feminist capacity to radically negate the very systematic and patriarchal space of identity through which the subaltern here speaks. Insofar as critical discussions of subaltern consciousness has obsessed over the modernity of the traditional, the praxis of the CMT illuminates a future egalitarian solidarity in present subaltern struggles.

Acknowledgment Note: I wish to express my deepest thanks to the members of the Community Media Trust, to my host and translator, Mr. Murthy, and to all the staff at the Deccan Development Society in both Pastapur and Hyderabad for their gracious assistance and cooperation. I also want to thank Dr. Gail Faurschou, Dr. Stephen Crocker, and Ashok Kumbamu for their assistance with carrying out this research. This research was funded by a research fellowship from the Indo-Canada Shastri Institute and a grant from the Social Sciences and Humanities Research Council of Canada. I thank them also for their support. Responsibility for any missteps remains my own.

REFERENCES

Bates, C. (1995). Race, caste and tribe in central India: The early origins of Indian anthropometry. In P. Robb (ed.), *The Concept of Race in South Asia* (pp. 219–259). Delhi: Oxford University Press.

Baviskar, A. (2006). The politics of being "Indigenous". In B. G. Karlsson & T. B. Subba (eds.), *Indigeneity in India* (pp. 33–50). London: Kegan Paul.

Chakrabarty, D. (2006). Politics unlimited: The global Adivasi and debates about the political. In B. G. Karlsson & T. B. Subba (eds.), *Indigeneity in India* (pp. 235–247). London: Kegan Paul.

Guha, S. (1999). *Environment and ethnicity in India, 1200–1991*. Cambridge: Cambridge University Press.

Hardiman, D. (1987) *The coming of the devi: Adivasi assertion in Western India*. Delhi: Oxford University Press.

Hardt, M. & A. Negri. (2000). *Empire*. Cambridge: Harvard University Press.

Innis, H. A. (1951). *The bias of communication*. Toronto: University of Toronto Press.

———. (1952). *The strategy of culture*. Toronto: University of Toronto Press.

Jameson, F. (2005). *Archeologies of the future: The desire called utopia and other science fictions*. London: Verso.

Karlsson, B. G. (2003). Anthropology and the "indigenous slot": Claims to and debates about indigenous peoples' status in India. *Critique of Anthropology, 23*(4): 403–423.

Kumbamu, A. (2006). Ecological modernization and the "gene revolution": The case study of Bt cotton in India. *Capitalism, Nature, Socialism, 17*(4), 7–31.

Patel, R. & A. R. Müller. (2004). *Shining India? Economic liberalization and rural poverty in the 1990s*. Policy Brief No. 10. Food First. Oakland, CA: Institute for Food and Development Policy.

Pinney, C. (1990). Colonial anthropology in the "laboratory of mankind". In C. Bayley (ed.), *The Raj: India and the British, 1600–1947* (pp. 252–263). London: National Portrait Gallery.

Shah, A. (2007). The dark side of indigeneity?: Indigenous people, rights and development in India. *History Compass, 5/6*, 1806–1832.

Skaria, A. (1997). Shades of wildness: Tribe, caste and gender in Western India. *Journal of Asian Studies, 56*, 726–745.

Wallerstein, I. (2004). *World-systems analysis: An introduction*. Durham, Duke University Press.

Xaxa, V. (1999). Tribes as indigenous people of India. *Economic and Political Weekly, 34*, 3589–3596.

CHAPTER 12

VOICING OUR ROOTS: A CRITICAL REVIEW OF INDIGENOUS MEDIA AND KNOWLEDGE IN BENGAL

SUDHANGSHU SEKHAR ROY AND
RAYYAN HASSAN

INTRODUCTION

Indigenous media in Bengal has been providing social, philosophical, and spiritual education to the people of the region since long before recorded history. In the past 200 years the greater Bengal region, which is comprised of Bangladesh and West Bengal of India, has been through three significant cultural shifts: the influence of British colonization; the partition and decolonization of Bengal into Eastern (Muslim majority) and Western (Hindu majority); and lastly, the liberation of Bangladesh as an independent state from Pakistan, in 1971. Throughout these tumultuous times the people of Bengal (both East and West) have managed to sustain their indigenous culture and traditional heritage amid a volatile background of political instability and foreign and domestic market-oriented cultural encroachment mechanisms. The significance of indigenous cultural processes in Bengal thus lies in their instrumental role in continually raising awareness among the rural people regarding various local issues, such as the protection of trees and natural ecological processes, conservation and use of water resources, and resolving social conflicts (Ghosh, 1996).

   This chapter examines the survival of indigenous media in Bengal within the sociocultural context of colonization and modern globalization.

Specifically, the chapter investigates the survival and growth of *Jatra* (folk theater) and *Baul gaan* (songs of Bauls) as examples of indigenous media. Kapoor (2009) argues that the problems faced by indigenous people of Asia include the plundering of resources, forced relocation, militarization, forced integration of indigenous peoples into market economies, and homogenization into a consumerist culture. Similarly, indigenous media of Bangladesh has been facing a growing threat via cultural globalization and associated homogenizations. A case in point is the use of the term globe*balai*zation rather than globalization by members of indigenous communities, where *balai* (evil or harm) is used to denote a cultural disease, which is corrupting the resident cultural heritage. The nation state of Bangladesh is composed primarily of an agrarian peasant society with a growing privatized upper middle class; however, indigenous media still manages to secure widespread and popular acceptability. This being said, there is a sense of growing modernization in the cities and expansion of markets, which is allowing foreign sounds and images, brands, and exotic labels to creep into the rural psyche. Furthermore, cultural pressures from the neighboring Indian film industry and associated high-voltage popular mainstream music and cable television are aggressively encroaching the traditional sensibilities of the culturally rooted rural Bengali society (Baten, 2008).

INDIGENOUS MEDIA IN BENGAL

Indigenous is that which is "born or produced naturally in a land or region" (Ahmed, 2000, p. 18). Indigenous media thus can be denoted as the cultural facets of an indigenous community such as its music, literature, art, and drama. In the case of indigenous media in Bengal, lack of modern technology in the rural areas leads the performer and the audience to interact face to face. However, in modern media, such as television, Internet, or radio broadcasting, a veil of impersonal technology has replaced direct communication between performers and audience. Moreover, repeated news broadcasts and visually appetizing images of celebrity high culture tend to leave audiences of modern media (namely radio and television) feeling disconnected.

Contrastingly, in the *ashar* (program or soiree) of *Baul gaan*, the singer and the audience sit in a close-knit circle without any distinction between them. The *ashar* can be perceived as somewhat of a group huddle, where people sit close together, at times conversing or debating human issues, and once in a while a Baul sings out a song that can initiate more dialogue or singing and keeps the overall interaction alive. This symbolizes that the performance or the song is not a separate imposition on the audience but rather a sharing of collective understanding among those present. Similarly, in the case of *Jatra* (folk theater), the audience is usually in very close proximity to the actors. The audience often applauds, comments on, or even criticizes the performers while the play continues.

As Mlama (1991) writes,

> The Popular Theatre process involved the participation of performers and whole communities in researching what the community felt were their problems. An analysis of the problems was done through interpersonal dialogue. The problems were then concretized into theoretical performances by the members of the community, using artistic forms which are generally familiar. A public performance was staged to present the problems and make suggestions for solutions. Eventually post-performance discussions by the performers and audience charted out what actions were to be taken by the community to solve the problems. (p. 5)

This aspect of popular theater is currently being employed by various NGOs (Non-Governmental Organizations) as a means of generating awareness among rural people regarding various development related issues such as health, education, and environmental consensus (ActionAid, 2004).

COLONIAL IMPACT ON INDIGENOUS MEDIA

In recent times the rural masses have constantly been exposed to media networks such as radio, television, and newspapers. The ideas being promoted through modern media are neo-liberal notions of "individual freedom, democracy, industrialization, urbanization growth of the Western world and economic growth" (Barua, 2009, p. 125). This contemporary cultural encroachment by the West is the offshoot of the European Christianization and consequent pressure of modernization.

Joseph (1996) argues that colonial theory heralded the myth of the white man's burden of civilizing the East from its savage and uncultured preindustrial backwardness. Therefore in the colonial rhetoric it was assumed that Europeans alone were capable of creating and expanding the notion of "modern civilization and culture" (p. 159). Hence, the "backward" Asians were required to be under the tutelage of colonial rule in order to be "civilized." Over the years, foreign mass media in Bengal has become glamorous and impersonal, whereas the familiarity of the performance of indigenous artists in the villages continued as local peasants "could not only see and hear but even touch" their performers (Joseph, 1996, p. 141). As colonialism probed deeper and deeper, the rural values of solidarity, spirituality, and harmony with nature were being forced to change into the individualistic values of self-centric materialism, competitiveness, and greed (Barua, 2009). This led to a feeling of subjugation and exploitation among the masses. As colonial sympathizers were growing into an elitist upper class, the disenchanted peasants began to reinvestigate their sociopolitical roles within their communities and villages. This is where indigenous media—art, literature, Jatra, and Baul music—became instruments of expressing resentment toward oppression and opened canvases for aspirations of liberation, unification, understanding, solidarity, and free-dom. Thus indigenous media in Bengal embodies the collective, creative activity of the whole people, and not the personal inspiration of a single ideology according to the monocultural system promoted by colonialists and the recent neoliberal consumerist rhetoric. This is exemplified in the various themes, legends, myths, imageries, and symbolisms that indigenous art and culture have abundantly drawn upon. Regardless of the adversity indigenous media faces from colonization and contemporary globalization, it still manages to secure its niche within the Bengali village.

Indigenous media in Bengal could be perceived as one of the key instru-ments in mobilizing the masses towards an anticolonial national movement. Throughout the history of Bengal there are many examples of playwrights, poets, or musicians who have been imprisoned or exiled for speaking out for the masses. For example, Mlama (1991) refers to Mukunda Das, a Bengali playwright, "who was sent to jail for his Swadeshi Jatra Plays which were directed against British colonialism" (p. 36). Similarly, Bengal takes pride in the likes of intellectual revolutionaries, such as Nishikanta Chatterjee (1852–1910), who was one of the pioneers of folk research in Greater Bengal and who earned his doctoral degree for his seminal work on Jatra in the 1880s. Mlama (1991) identifies the following verse as it aptly describes the Bengali people's rhetoric during the colonial period:

> The flag of revolution, the call to insurrection,
> Listen to the toilers' cry!
> For ending, injustice, destroying oppression
> The moment of truth is nigh-
> Come, come you poor, come, take the flag in hand,
> Listen to the toilers' cry
> Moneylenders' rule cannot be endured,
> Now it must be destroyed!
>
> Mlama, 1991, p. 35

ORIGINS OF INDIGENOUS THEATER

Ahmed (2002) identifies Arthur MacDonnell, the author of *A History of Sanskrit Literature* (1917), who describes the source or origin of dramas of ancient India to be prior to the twelfth century. Bagchi (2007) identifies the dramatist, writer, and actor Mamtazuddin Ahmed who also suggested that the open-air theatre might have been popular before the Vedic age predating the twelfth century.

Since at least the twelfth century then, the indigenous theater heritage of Bengal has existed. It has developed into distinct forms including Sanskrit *Natyakala* (Sanskrit Fine Arts), *Naat-Geet* (Classical music), and puppetry. National Professor of Bangladesh Kabir Chowdhury confirms that the Jatra is rooted in the soil and people of Bengal (Mahmud, 2008), highlighting the long history of this indigenous medium.

MEANING OF JATRA

Literally, Jatra means "going" or "journey." It is a form of folk drama combined with acting, songs, music, and dance, and is characterized by its stylish delivery and exaggerated gestures and orations. Jatra is believed to have developed from ceremonial functions held before starting on a journey. In his doctoral thesis, *The Yatras or The Popular Dramas of Bengal* (1882), Chatterjee writes that the word Yatra is derived from the root ya-, which means "to go". Yatra therefore means, in the first place, a going, a departing; for example, *Usha-yatra* (leaving home at the earliest dawn); *Maha-yatra,* the great departure—that is, death—hence a pilgrimage (Asghar, 2002).

JATRA IN THE CONTEXT OF THE PARTITION

Jatra was instrumental in opposing the colonial promotion of religious difference during the 1905 partition of Bengal. A movement popularly called the Swadeshi Movement was waged under the leadership of Indian Congress leader, Ashwinikumar Dutta (1856–1923) and his associate, Mukunda Das (1878–1934). The Swadeshi Movement was successful in reversing the partition of Bengal but failed to win popular support of the Muslims, evident in the use of cultural symbols such as *Vande Mataram,* which was the national cry for freedom from British rule during the freedom movement (Chakrabarty, 1997). Sir Herbert Risley, one of the chief architects of partition, stated, "One of our main objects is to split up and thereby weaken a solid body of opponents to our rule" (Ahmed, 2002, p. 288).

In the face of such colonial forces, the Swadeshi Movement in general, and the Swadeshi Jatra in particular, integrated both Hindu and Muslim sentiments of liberation and did not espouse any separatist communalism. This notion can also be observed in some of the songs of Mukunda Das, such as, *"Rama Rahim na juda karo bhai"* [Don't differentiate between Hindus and Muslims] or *"Sajre santan Hindu Mussalman"* [Get both the Hindu and Muslim children ready] (Ahmed, 2002, p. 88).

However, since the partition of Bengal, the present development-culture clash and the implication of globalization and emerging religious fanaticism have threatened the survival of Jatra (Ahmed, 2002). The tenuous relationship between development and culture has been aptly identified by Mlama (1999) who states that:

> Clashes between . . . economic and or technological imposition and the indigenous cultures had led to the accusation that culture impedes development.

> Indeed, indigenous cultures have often been labeled backward and anti-development. The assumption is that any society wishing to develop needs to discard its indigenous cultures. (p. 16)

There have been instances where Jatra artists had been forced into other professions as the usage of the indigenous medium was banned through repressive laws introduced by the British colonial rulers in the name of public order, security, and public administration (e.g., Bengal Places of Amusement Act 1933). As Joseph (1996) writes, "[f]olk culture is spontaneous, collectively created and shared and is created by people who are in closer association with nature" (Joseph, 1996, p. 152). Herein lies the strength of folk culture like Jatra or Baul songs.

FORMS AND VARIETIES OF JATRA

Brindavan Das, the author of *Chaitanyabhagavad* (1548), a Hindu Holy scripture, described a dramatic performance during which Shree Chaitanya (1486–1533), a Hindu *avatar* (incarnate) performed the role of Rukmini, the wife of Lord Krishna. Scholars such as Kapila Vatsayan believe this performance to be the birth of *Krishna Jatra*. With the development of Vaisnavism (a concept of being isolated from the earthy world), Krishna Jatra spread through Bengal. Unlike Western drama, there was no dramatic conflict in Krishna Jatra; rather it was confined to one of the nine classical *rasas* (sentiments), the *shringar* (erotic). Krishna Jatra emerged as one of the leading performance genres in the seventeenth century. By the eighteenth century, a number of other forms of Jatra had developed: *Shakti Jatra, Nath Jatra, Ram Jatra,* and *Pala Jatra.* Krishna Jatra and Chaitanya Jatra, however, continued to dominate. In the eighteenth century, Jatra flourished in Vishunupur, Burdwan, Beerbhum, and Nadia in the West Bengal region of India and in Jessore of Bangladesh (Ahmed, 1997).

JATRA AS A NATIONALISTIC MOVEMENT

Jatra fostered a strong sense of patriotism among the masses. With the help of myths and legends, it so effectively promoted an anti-imperialist sentiment that many performances were prohibited (Ghosh, 1996). A major change in Jatra took place after the First World War when nationalistic and patriotic themes became incorporated into the Jatra. Though religious myths and sentimental romances continued to inspire the Jatra, the nationalistic and patriotic spirit of Bengal also found its expression in the Jatra. Mukunda Das and his troupe, the Swadeshi Jatra Party, performed Jatras about issues such as: colonial exploitation, patriotism, anti-colonial struggle, and oppression of the feudal and caste system. In the 1940s, when the struggle for independence from colonial rule was nearing its climax, the sociopolitical content of Jatra superseded the religious-mythical theme.

Jatra as a social agent was (and is still) being played in Bengal to create dialogue for socialist solutions. As a result various Jatra troupes have been composing and presenting different *palas* (episodes) to the masses of Bengal that are based on the lives of renowned individuals, such as: Vietnamese leader Ho Chi Minh, Soviet leader Vladimir Lenin, German autocrat Hitler, anti-imperialist Bengal Nawab Sirajuddowlah, and poet Michael Madhusudan Dutta. The intention of these *palas* is to instill a sense of liberation and hope in the subjugated classes. A renowned scholar of East Bengal, Ibrahim Khan once took a *Moulvi* (an orthodox Muslim cleric) of Noakhali, which is a conservative region of eastern Bangladesh, to a *Jatrapala* (episode). Initially the Moulvi thought that enjoying Jatra was a sin, but after witnessing a Jatrapala presented by Mukunda Das, the Moulvi changed his attitude. The Moulvi was curious about the manner and form through which Mukunda Das focused on patriotism. The Moulvi said, "for extending services for the humanity and directing them to go on in the right path towards freedom what Mukunda Das said is truly a narration of verses from the holy books" (Bagchi, 2007, p. 128).

Jatra has tremendous power to influence people to unify. In spite of facing a long-handed onslaught of globalization and Western modernization, Jatra is still successful in uniting people especially during Bengali Hindus' greatest religious festival *Durga Puja* in October, and during the winter seasons when peasants enjoy the luxury of time after harvesting. In Bangladesh during the *Durga Puja* people from different remote villages arrange consecutive *Jatra Palas* (episodes) for five consecutive nights, which draw huge audiences of both sexes. The pioneering poet of Bangladesh, Shamsur Rahman (1929–2007), accordingly said, "Jatra even today stirs the midnights of thousands of audiences, and sprays earthy good feelings in public mind" (Bagchi, 2007, p. 54).

LALON SHAH AND THE BAULS OF BANGLADESH

Another important form of indigenous media in Bengal is *Baul Gaan,* or the Songs of Bauls. The term *Baul* literally means the seeker of air or breath in Bengali. Although *Baul Gaan* have been sustained for an estimated thousand years in the Greater Bengal region, the period of its true origin is still a matter of debate. The first textual evidence dates back to the fifteenth century; however, according to local interpretations, it is claimed to be somewhere in the eighth century. The spirit of Baul philosophy is a mingling of Sufism, Vaishnavism, and Buddhist Tantrism to explain the human body and soul. The philosophy highlights humanism and metaphysical relationships between the Creator and the creations. It is this mystic notion that still sweeps across the emotional states of the people of central and western regions of Bangladesh, and southeast and northern parts of West Bengal.

The emergence of Baul philosophy as a cultural social movement cannot be accredited to one single person. However, the Saint Fakir Lalon Shah from Kushtia is one of its greatest exponents. Lalon Shah was a disciple of another mystical poet, Shiraj Shain. This *paramapara* or relationship between the master and the disciple is the only means through which the Baul philosophy disseminates esoteric knowledge among its members and thus ensures the security and survival of Baul cultural identity. Baul philosophy indoctrinates the method of master-disciple knowledge transfer from three distinct religious practices namely, *Krishna Bhakti,*[1] Sufi tariqah,[2] and *Zen Buddhism.*[3] Even though the secrecy maintained between the master and the disciple ensures the survival of the Baul philosophy, this very notion leads to a dearth in the amount and quality of documented information regarding the Bauls (James, 1987).

Wandering mystics, also known as fakirs,[4] and Bauls existed and traveled long before the emergence of Lalon Shah. Due to his acceptance among the masses, many Bauls and fakirs came to him seeking knowledge and guidance. It can be posited that Lalon Fakir, as he was sometimes called, was instrumental in integrating the fakirs and the Bauls into a common philosophical grounding. He was born presumably in 1774 AD[5] at Harishpur village in the Jessore district of Bangladesh. He later migrated to Cheuria village in the district of Kushtia in about 1804 where he spent his remaining years.

Lalon Shah's community was composed of a people who had left behind their material possessions in a higher pursuit of understanding of existence. Due to their impoverished appearance and ascetic lifestyle, they were often depicted as strange by mainstream society. Yet regardless of their eclectic demeanor they have managed to secure a stoic presence in Bengali culture and history. In the modern age the Bauls have survived the social influence of British colonization, the Urdu influence of post partition of East Pakistan, and finally the independence of Bangladesh as a nation state in 1971. Throughout these three shifts, Lalon Shah's doctrine and Baul philosophy have not only survived but have gained acceptance among all classes and races of people independent of religions in Bengal.

BAUL MUSIC

Baul performances are often considered as sacred and mystical experiences. Through the song, the singer and the listener join in a common understanding of belonging to the land and greater humanity. The Bauls usually travel from village to village spreading their music and message among the rural households. According to the Bauls, a person has to look inside oneself to comprehend the idea of divinity. All over Bengal the Bauls agree that all the knowledge and the secrets reside within the human body. Furthermore the Baul philosophy proposes that through *sadhana,*[6] dance, and song, one can achieve the liberation of the human soul and thereby open the pathway for Divine knowledge. This has remained unchanged throughout history. As a result the Baul movement has been able to adapt to mystic philosophies in the rural value system even amidst the changing cultural phases in the Greater Bengal region; thus it has survived. Baul songs are usually solo songs performed in *ashars* (soiree), although often members of the audience join in the refrain and chorus of the verses. Some of the instruments used in *Baul ashars* include:

- *Ektara*—a plucked single string drone, where fingers and thumbs are used in a flicking motion to produce single note rhythms;
- *Khamak*—a rhythmic instrument with one or two strings attached to the head of a small drum. The strings are plucked with a plectrum and they are alternatively tightened or slackened to generate an amazing array of rhythmic variations; and
- *Khol*—A barrel shaped clay drum, generating a simple progression of percussion sounds (James, 1987, p. 17).

Mysticism has been an integral part of the Bengali psyche. As Haroonuzzaman (2008) observes, "[h]owever modern we claim to be, every Bengali worth his or her salt feels [that] the pull of the roots and the quest for something beyond the material world is always present" (p. 4). Baul music with its insightful look into the notions of self, social class paradox, and freedom provides the individual with cultural root knowledge, which seems to be lacking in most mainstream media. However, it is significant that nowadays, regionally popular musical groups, such as *Bangla* and *Leela,* have brought folk music and especially Baul music to the mainstream modern Bengali audience (Reverbnation, 2009).

THE MESSAGE OF LALON SHAH

Lalon Shah used the Bengali language as the principle medium of information exchange. Bengal's cultural history has faced the influence of other languages such as Sanskrit, English, Urdu, and Hindi. However the use of the mother tongue, Bangla (the Bengali language) enabled Lalon Shah's message to reach the masses. According to the local people of Kushtia, Lalon Shah composed over 9000 songs in his lifetime, however documented statistics regarding these figures have been almost impossible to locate due to the nature of oral transfer of knowledge between the master and the disciples. Due to the esoteric nature of Baul music where notions of spirituality, transcendence, and harmony with nature are issues that intermingle, it becomes difficult to bracket these metaphysical themes into a singe form or type of music. Regardless, Lalon Shah's body of work, commonly called *Lalon Geeti* in traditional Bengal, reveals that Baul music and songs can be placed under several principal themes or *totto*:

Allah totto—Knowledge of the Divine; Rasul totto—Knowledge of the Prophets; Sristi totto—Knowledge of Creation; Murshid totto—Knowledge of the Masters; Atta totto—Knowledge of the Soul; Deho totto—Knowledge of the Body; Manush totto—Knowledge of Humanity; Mon totto—Knowledge of the Mind or Intellect; Porom totto—Knowledge of the Secrets; Jati totto—Knowledge of the Races; Parapar totto—Knowledge of the Way. (James, 1987, p. 37)

Pilgrims and travelers from the three main religions Buddhism, Vaishnavism, and Islam, were often passing through Kushtia. Even now there are various stories prevailing about Lalon Shah's meetings and discussions with *Sadhus*[7] and *Sufis*[8] in Kushtia. These exchanges in dialogue with different sects and cultures made a deep influence on Lalon Shah's music and philosophy. Due to his experiences with these diverse philosophies, Lalon Shah ascribes to a holistic view of humanity. This is best represented in one of his songs where Lalon Shah says: "Shobai bole Lalon ki jat shongshare, Lalon bole jatir ki roop dekhlam na jibon ta te" (James, 1987, p. 32). [Everybody keeps asking who is this Lalon, what is his caste or creed in society? Lalon says in my lifetime I have yet to witness what color or creed this society has].

During the end of the colonial period, there was a cultural paradox deep within the subcontinent. Hindu and Muslim separatists prepartition began to operate mutually exclusively whereas the great British Empire was the patriarchal figure who would be an instrument of conflict resolution and play favorites according to their interests. This larger drama led to microcosmic religious conflicts along with feudal class differences in the rural areas leaving behind many needlessly dead and the remaining disillusioned and confused (Ahmed, 1997).

It is at this juncture where Lalon Shah's music comes to the people of this region. His music cut through the barriers of religious labels and political propaganda. The political wave at the time required people to side with Hinduism, Islam, or British Colonial subjugation. Here, the innocuous peace loving Bauls appear as a strange misfit in the social order of colonial conflicts, with their long flowing hair, their simple one-piece clothing, and their knowledge of fasting, singing, meditation, and communal harmony. Amidst a backdrop of racial and religious intolerance, corruption, and rampant violence, the Bauls were a welcome panacea to the beaten and defeated masses. As the political ramifications of partition began to surface, complexities and inequities started emerging regarding core issues such as land ownership, taxation, and citizenship; the Bauls on the other hand were setting an example of living a life of sheer simplicity.

The notion of *Sheba*—the receiving and sharing of food in equal portions among all Baul members—was a gesture binding the Bengali humanity in a time when Bengal was being divided into many parts East, West, Hindu, Muslim, Buddhist, Christian, Communist, and Company (British East India Company). In that rancorous environment of political propaganda and racial rhetoric the Bauls spent most of their time in silence away from society. Artful methods of breathing were taught and practiced which would aid the body in curbing negative energies such as desires of greed, lust, jealousy, and power in exchange for peace, unconditional love, and harmony with nature. The use of simple agricultural produce for instruments, soulful lyrics, and an uncompromising sense of expecting nothing in return other than what comes to the self by natural law were just some of the teachings of Lalon Shah, and the Bauls have continued to follow his footsteps.

THE SACRED TEACHINGS OF LALON SHAH AND NATURAL LAW

Lalon Shah's philosophy allows people of all walks of life to find a sense of meaning in being true to their independent natural selves. From an academic perspective this prohumanity philosophy is common to the Western notion of human rights. For example, Brown (2005) suggests that the case of human rights is built on the foundation or idea of natural order. He adds that universal moral standards have existed and the rights of each individual adhere to these same standards. This position of individual rights is thus not limited in application to any particular legal framework, community, race, state, creed, or civilization (Baylis & Smith, 2005).

Lalon Shah's role as one of the common folk singing outside the boundaries of religious labels, political interests, and even the much harder box of economic laws of market and labor, depicts him as an icon of Bengali rural identity. His lifestyle as an ascetic vegetarian meant that his dietary requirements barely taxed the environment. Disciples of Lalon Shah strictly follow the monastic discipline even now, where the ethical core value still remains the same—a total submission to nature. To this end Lalon Shah's philosophy of being true to nature must be understood in the context of harmony and discipline and not a view of wild abandon of animal instincts without any boundaries. It can be thus posited that the modern notion of dominating nature by the concept of industrialization, mass production, and "free" enterprise in a world without consequence of action is foolhardy.

Lalon Shah's songs used simple and straightforward lyrics. Without any pretense in his message his aim was not to target the community or the group but rather the forgotten and misled individual. The following verse exemplifies Lalon's focus on the individual:

> Boro shonkote poriya doyal bare bare daki tomai
> Khomo, khomo khomo oporadh
> Rakho Maro shey nam Nobi'r
> Ami ki tor keho noi?
> Khomo Khomo Oporadh
>
> James, 1987, p. 45

[In great difficulty I keep knocking on your door O Great Creator of Being
I beg forgiveness for my sin

In life as in death is the name of the Prophet (pbuh)
Then what of me, am I nothing to you?
I beg forgiveness
I beg forgiveness for this insolence]

The people of Bengal have kept Lalon Shah's work alive in their hearts for over a century. His music is being recorded by a new generation of musicians using modern instruments. Although the sounds of his songs and his audience have changed, the message of unity, compassion, and unconditional love and acceptance across all peoples remains the same.

AN INDIGENOUS CULTURAL MOVEMENT CONTINUES

The significance of Jatra and Baul music has taken a step from the rural society into the modern social atmosphere inspiring people's movements for equality and justice throughout Bengal. Examples, such as the Tebhaga Movement in 1947, which was organized by the Adivasi Santal community; and the great Language Movement in 1952 against the Pakistan government authority, are both sociopolitical movements that have been driven on by the cultural roots of Bengal. UNESCO has officially recognized February 21 as International Mother Language Day since the language movement of 1952 in Bengal. At present, sociocultural movements, such as the Gram Theatre Movement initiated by recently demised dramatist-researcher Selim-Al-Deen, are continually promoting the voices of the masses. A noted scholar and Professor of the Dhaka University, Sirajul Islam Chowdhury, once said,

> Colonialism or what we might say imperialism now, has capitalism at its core. As capitalism is a system that sees everything, even culture, as a mere product, any cultural activity cannot but move forward is direct conflict with capitalism. . . . Colonialism works, by importing its culture as a superior one. In order to counter this, we must cherish our own cultural heritage, which is indeed, a good deal richer than that of western culture.
>
> (Sarwat, 2008, p. 9)

UNESCO has also proclaimed that the traditional Baul songs of Bangladesh are one of the forty-three masterpieces of oral and intangible world heritage (Kamol, 2008). Recently the Department of Theatre and Music of the University of Dhaka included *Jatra pala* (episodes) in its formal curriculum. From now on, third year students of the department will study the medium as one of their courses. Dr. Israfil Shahin, the Chair of the Department said,

> Through the inclusion of Jatrapala in our curriculum, the department focuses on the vast treasure of indigenous art forms of Bangladesh . . . this initiative will help students to delve deeper into our heritage, roots . . . Jatrapala is a cultural archetype and initiative to regain its lost glory is the need of the time.
>
> Mahmud, 2008, p. 8

Every year, hundreds of thousands of followers of Lalon Shah throng to Cheuria to commemorate the anniversary of his death. In 2008 a massive gathering of followers was present during the death anniversary, mainly as a protest against the demolition of a sculpture of Lalon Shah in front of the Dhaka International Airport by orthodox clerical fanatics. This catastrophe touched the secular sensitivities of the nation. As a consequence, rallies were organized by artists, singers, academics, intellectuals, and cultural activists at all cultural centers and private and public universities in the country, once again proving the power and sustainability of folk media in Bengal.

CONCLUDING REFLECTIONS

This paper described the intricate nature and form of indigenous culture and traditional heritage throughout the vast area of Bengal. The paper further investigated the cultural components that are interlinked with indigenous media. The metamorphosis of indigenous media as a voice against imperialism was elaborated through the examples of various playwrights, artists, and cultural activists. The nature of their work was principal in resisting undesired transformation of the Bengali culture. The historical context of the indigenous theater, and sociocultural viewpoint of the development of its numerous forms were also discussed. Emphasis was given to the pivotal role of indigenous media (especially Jatra and Baul Music) to the movement against colonization and establishing the long existing cultural framework of Bengali identity. The art form of Jatra was presented as a major social agent in launching national agenda, and at the same time as a full-fledged medium to voice against exploitation and foreign dominance.

In a similar vein the paper delves into the unique form of Baul music and its stoic acceptance among all classes of Bengalis. Baul music has been depicted to carry expressions of immense depth and knowledge about the nature of Bengali mysticism and consequently it maintains a profound effect on the minds of listeners even today. This shows the timelessness of the messages in Fakir Lalon Shah's songs. The relationships of the human, body, and soul and the sense of spirituality and harmony with nature were also shown to bring about a solidarity and understanding among Bengalis from diverse cultures and religions. Furthermore, the paper looked into the significance of the Bengali language as a unifying factor for the diverse population of its people. The significance of the Bengali language as the principal medium in both Baul music and Jatra performances was a key element in mobilizing the peasants and dissolving boundaries of race, creed, class, and religion. The paper concluded by identifying the nature of contemporary indigenous movements such as the Gram theater movement and the practice of cultural art forms in different educational institutions and cultural platforms both at the local and national level. The recognition of International Mother Language day and Baul music as an example of intangible world heritage exemplify the significance of Bengali indigenous media and its importance not only from a regional but also from a universal perspective. As this paper is being written *Jatra, Tarja, Tamasha, Kobigaan, Gambhira, Tamsha, Baul Gaan* are all still being eloquently practiced in their diverse forms and colors across the microcosmic universe of villages and riverbanks of Bengal.

Bengali indigenous media still continue to inspire and unite the peasants, farmers, fishermen, and middle class alike amid the chaos of modern cultural encroachment and crippling market dominance. This paper thus aims to present the Bengali people not just as a group of disgruntled victims of postcolonial global hegemony, but as one of the many unrecognized and rare cultures who sit under the stars and practice their unique form of music and art, forever probing into the deeper mysteries of the self, the planet, the universe, and everything else it contains.

REFERENCES

ActionAid (2004). *To enter again the sweet forest: A qualitative perspective of livelihood change in the Sundarbans.* Dhaka: Intent Publishers.

Ahmed, J. S. (1997). Drama and theatre. In S. Islam (ed.), *History of Bangladesh 1704–1971: Social and cultural history (Vol. 3)* (pp. 473–542). Dhaka: Asiatic Society of Bangladesh.

———. (2000). *Acinpakhi infinity indigenous theatre of Bangladesh.* Dhaka: University Press Limited.

———. (2002). Bengali nationalism through sociology of theatre. In A. M. Chowdhury & F. Alam (eds.), *Bangladesh on the threshold of the twenty-first century* (pp. 286–306). Dhaka: Asiatic Society of Bangladesh.

Ahmed, Momtazuddin (2000). Practice of Jatra: The crisis and development. In S. M. Mohsin (ed.), *Jatra our heritage* (n.p.). Dhaka: Bangladesh Shilpakala Academy.

Ahmed, Moyeenuddin (2002). *Bangladesher Jatra Ebong Prasnaga* (Banglar Loksangskriti: Jatra Shilpa). [Jatra of Bangladesh and Context (Folk culture of Bangla: Jatra industry)]. Dhaka: Bangla Academy.

Asghar, S. (ed.) (2002). *Banglar Loksangskriti: Jatra Shilpa.* [Folk culture of Banga: Jatra industry]. Dhaka: Folklore Division, Bangla Academy.

Bagchi, T. (2007). *Bangladesher Jatragaan: Janomdhyam O Samajik Pariprekshit.* [The Yatras of Bangladesh: Media and social context]. Dhaka: Bangla Academy.

Barua, B. P. (2009). Non-formal education, economic growth and development: Challenges for rural Buddhists in Bangladesh. In A. Abdi & D. Kapoor (eds.), *Global perspectives in adult education* (pp. 125–140). New York: Palgrave Macmillan.

Baten, A. S. (2008). *Globalization and anti-globalization.* Dhaka: Pathak Shamabesh.

Baylis & Smith (eds.) (2005). *The globalization of world politics: An introduction to international relations* (3rd ed). New York: Oxford University Press.

Chakrabarty, B. (1997). *Local politics and Indian nationalism: Midnapur (1919–1944)*. New Delhi: Manohar.

Ghosh, S. (1996). *Mass communication today in the Indian context*. Calcutta: Profile Publishers.

Haroonuzzaman, M. D. (2008). *Lalon: Selected Lalon songs translated into English*. Dhaka: Adorn Publications.

James, B. (1987). *Songs of Lalon*. Motijheel: University Press Limited.

Joseph, M. K. (1996). *Modern media and communication*. New Delhi: Anmol Publication.

Kapoor, D. (2009). Globalization, dispossession, and subaltern social movement (SSM) learning in the South. In A. Abdi & D. Kapoor (eds.), *Global perspectives on adult education* (pp. 71–92). New York: Palgrave Macmillan.

Kamol, E. (2008, Oct. 16). Controversies shroud Lalon and his songs. *The Daily Star* (Dhaka), p. 9.

Mahmud, J. (2008, Nov. 13). A new hope for indigenous performing arts. *The Daily Star* (Dhaka), p. 9.

Mlama, P. M. (1991). *Culture and development: The popular theatre approach in Africa*. Uppsala, Sweden: Scandinavian Institute of African Studies.

Reverbnation (2009). *Bangla music—Reverb nation*. Retrieved from, www.reverbnation.com/bangla

Sarwat, N. (2008, Nov. 16). Bangladesh gram theatre convention. *The Daily Star* (Dhaka), p. 8.

PART IV

GENDER, INDIGENOUS KNOWLEDGE, AND LEARNING

CHAPTER 13

HAYA WOMEN'S KNOWLEDGE AND LEARNING: ADDRESSING LAND ESTRANGEMENT IN TANZANIA

CHRISTINE MHINA

INTRODUCTION

Literature on African women's marginalization has expanded over the past three decades. Much as critical voices both in the academy and in local communities are drawing attention to the fact that subordinated people's knowledges have been left out of academic texts and discourses, there has been little discussion on how women can utilize their knowledge to address issues of their concern. This chapter illustrates the potential of indigenous women to know, to choose, to imagine, to create, to decide, and to act. I use the Tanzanian case to show local women's creativity and resourcefulness to develop indigenously informed collective solutions to address their marginalized rights to land tenure.

This chapter highlights the women's marginalized position within the existing Haya indigenous land tenure system and how the colonial state impositions have further marginalized Haya women in relation to their access and control of land. Based on my participatory research initiatives with the Haya community, undertaken at Maruku village in Bukoba District of the Kagera Region in Tanzania between February and September 2002, the chapter illustrates how Haya women activated and applied their own traditional knowledge to deal with this situation. The PAR (participatory action research) process involved a group of ten women: six widows, two married women, one divorcee, and one single unmarried woman, who all lived in Maruku village. Because they all knew each other as members of Maruku village, there was an environment of mutuality and comfort during the PAR process, which encouraged them to share their personal experiences. The chapter concludes with a few reflections on the prospects for change with regards to addressing land discrimination of Haya women.

The Haya are original inhabitants of Kagera Region of northwestern Tanzania. Subsistence lifestyle, patrilineal inheritance, cultural norms of clan exogamy, and values of collectivity define the Haya ethnic group. A common set of social norms not only defines appropriate and inappropriate values, beliefs, and behaviors, but also acts as indicator of contrast to other ethnic groups of Kagera region. Social organization in Haya society is profoundly circumscribed by patrilineal kinship norms, which divide up the society and place members of families into clans. People in Haya society are mostly smallholder farmers who tend small patches of land, mainly cultivating banana plants, beans, maize, and root plants, as well as cash crops like coffee, tea, cotton, and sugarcane. Some people keep a few cows for milk and manure and some have goats and chickens.

Land is the most basic resource of agricultural production in Haya communities as it provides the production possibilities for growing food and other crops, as well as fodder for livestock. In addition to its immediate economic importance, land provides an avenue for social affiliations among members of a Haya clan as it structures people's relationships with others (Migot-Adholla & John, 1994). The majority of farmers in Haya society utilize land held under rights of clan ownership (Mhina 2005, 2009). The right to cultivate is generally predicated on membership in land-controlling lineage through birth and marriage. In most cases initial rights to use particular parcels of land is derived from first the clearance of virgin territory by a male family member, usually the family head, who then allocated plots within the area. This parcel of land, which will be the permanent homestead for the lineage and which also represents the clan land, is known as the *kibanja*. While different authors use different names for these customary arrangements of landholding, I prefer to use "indigenous land tenure system."

In this indigenous land tenure system, land as a resource is inherited patrilineally. Based on principles of clan exogamy (the prescribed practice of marriage outside the kin group, the boundaries of which are defined by the incest taboo) and patrilocal residence, the wife moves into her husband's clan territory after marriage (Lugalla & Emmelin, 1999). Thus women typically lose use rights over land in their natal lineages after marriage. Women rarely inherit land in their capacity as adults or wives. Adult women with children but without a spouse (de jure and de facto female heads of households) face the greatest difficulty in accessing land. Today, with the increasing incidence of unwed mothers, widows, and divorcees this problem is even more widespread today that it has been in the past.

As mentioned above, an indigenous land tenure system is a socially embedded system. For Haya society it is a rule-governed arrangement that is characterized by informal local-level customary practices based on social norms of Haya people (Migot-Adholla & John, 1994). Thus, claims to use and dispose of land arise out of social relations, that is relations between people rather than out of property relations, which are relations between people and things. Thus, the indigenous land tenure system in Haya society is a body of knowledge generally derived from long established patterns of behavior and practices within the Haya social setting. It is a sort of people's "law" that is based on the values and norms expressed in the customs and traditions of Haya people.

Control over land was a key area of struggle between the colonial state and the kinship/chieftainship-based political institutions, particularly between 1890 and 1930 (Mamdani, 1996). History reminds us how colonists imposed the alien hierarchical Western knowledge system as an attempt to replace indigenous knowledge systems (Toulmin & Quan, 2000). With regard to land tenure, colonists projected their own (European) models promoting individualized landholdings, for example with a freehold tenure system and a court system for land dispute resolution. New concepts of land tenure, for example usufruct rights, land ownership, and customary land law, which were based on ethnocentric and ideological biases onto African societies (Purcell, 1998) were introduced and applied in court systems. Stereotypes of indigenous tenure systems often bear little resemblance to reality, however, these systems were judged "imperfect" by colonial officials and elites, in comparison to idealized Western property institutions (Bruce, 1993).

The process of Westernizing indigenous land tenure interfered with the existing social rules and norms and actually exacerbated indigenous women's already weak position. Early anthropologists documented how colonists attempted to systematize indigenous practices of all African land-holding patterns into a framework of institutionalized settlement procedures, named customary law (Migot-Adholla & John, 1994; Peters, 2004). However, it is claimed that what was written as customary law was actually colonial rulers' misinterpretations of the open and flexible precolonial patterns of land use. Thus, European biases and stereotypes of African customs became encoded in new but incorrect definitions of indigenous tenure systems. In some instances, the formalization of indigenous land tenure systems strengthened men's individual ownership by giving the man the right to sell the land. This problem was compounded by the complexity of the court system with its interrelated legal texts, which were unfamiliar and poorly understood even by members of the local communities (Migot-Adholla & John, 1994). Thus, women lost the customary claims they previously held, weak as they may have been.

What is evident here is the very different ways of constructing and thus legitimizing knowledge. While the Haya people in their communities conceptualize their indigenous land tenure system in terms of relations among people, the westernized notion of land tenure perceives land as a property to be bought and owned, which distinguishes those who have the capacity to buy and own land as superior. Although the social patterns and arrangements in indigenous land tenure do not correspond to the Western conceptions of ownership and usufruct rights, policy makers at the national level, who have been trained in colonial educational systems, still repeat the same colonial mistakes of replacing the indigenous institutions with Western models of land policies and the preferred court system for all land disputes. What is not debated is the fact that these policy makers, whose lives are spent in one type of milieu may know very little about how others in less privileged sectors live and they do not know the problems that are encountered with the newly introduced court system. Based on the results of my research with a local community (Mhina, 2005), I learned about various difficulties with the court system from the women as they shared their experiences in a learning circle (described later in the chapter). I argue that indigenous people, including the Haya women, are highly knowledgeable about their social cultural environment and their problematic situation and indeed more so than outsiders. This chapter illustrates how people in indigenous communities, in particular Haya women, can work collaboratively to improve the existing indigenous land tenure systems by changing those aspects of the systems that disadvantage women rather than replacing them with imposed and unfamiliar Western models.

There is abundant evidence that indigenous knowledge systems throughout Africa are the bedrock of indigenous communities' survival (Dei, Hall, & Rosenberg, 2000; Mhina, 2005, 2009; Mhina & Abdi, 2009; Mikell, 1997; Semali & Kincheloe, 1999; Shizha, 2009). Not only that, in various African cultures women are regarded as custodians of tradition and culture and they have the responsibility of passing it from one generation to another. Needless to say, people in indigenous communities know what is valuable knowledge and what is not (Mikell, 1997); they know from their own life experiences through trial and error not only how to treat diseases, tend livestock, manage aquatic resources, and provide health therapies, but also how to preserve and pass on such local knowledge from one generation to the next (Semali, 1999; Shizha, 2009). What this illustrates is not only the significance of indigenous knowledge for the survival of indigenous people in their communities but the idea that women in indigenous communities are useful community educators.

HAYA WOMEN AS INSTIGATORS OF SOCIAL TRANSFORMATION

My PAR with a Haya community was motivated by a desire to explore the potential of local women as community educators. Specifically, to what extent can local women use their resources and apply their knowledge to begin to change their marginalized rights to agricultural land? As Dei et al. (2000) pointed out, indigenous people's experiential reality of the world reflects the capabilities, priorities, and values systems of their communities. Based on various African scholars and critics of African development (e.g., Dei et al., 2000; Mhina & Abdi, 2009; Semali, 1999; Shizha, 2009), my own research has developed on the assumption that any meaningful development could only be achieved if the people's culture and popular knowledge are integrated into the development process (Mhina, 2005). It is in a similar vein that Semali and Kincheloe (1999) share their strong belief in the transformative power of indigenous knowledge and the ways that such knowledge can be used to foster empowerment and justice in a variety of cultural contexts.

Following from this, I took the initiative to engage in a PAR that sought to break the silence around the knowledge held by Haya women. The PAR process in this context was aimed at involving the oppressed and disenfranchised local women in the collective investigation of their reality in order to transform it (Hall, 1993; Mhina, 2005). It was approached as an emancipatory approach to knowledge production and utilization. The Haya women actively took part in defining their social and cultural realities within their cultural settings. Although PAR, as a methodology, has evolved and transformed considerably since its beginning more than fifty years ago, even now, it continues to be the most effective form of research for working with marginalized communities and populations across a wide range of social and geographical contexts (Kapoor & Jordan, 2009).

ACTIVATING HAYA WOMEN'S KNOWLEDGE THROUGH A WOMEN'S LEARNING CIRCLE

I drew insights from Freire's (1973) work in choosing PAR as the methodology to lead to meaningful answers in exploring to what extent women can extend their experiential knowledge to address issues related to their land tenure discrimination. Freire (1973) believed that the goal of progressive educators should be to help men and women help themselves, by placing them in consciously critical confrontation with their problems and making them agents of their own recuperation. In other words solutions to problems need to be found with the people and not for them or superimposed upon them. In his discussion on communication and extension of technical knowledge between agronomists and peasant farmers, Freire (1973) argued that "scientific" knowledge becomes scientific as it takes on board the knowledge of the people. Furthermore, the knowledge of the people becomes knowledge for action and effective change when it takes on board in a creative way the "scientific" knowledge offered by intellectuals. Likewise, I viewed that my task, as a participatory action researcher, was to challenge women to penetrate the reality they are confronted with; that is, their marginality in terms of their rights to land and the subjugation of their position as community educators. In this regard, I agree with Rosenberg (2000) that in studying indigenous knowledges, we need to approach marginalized indigenous groups in the context of their own experiences and histories, with the goal of centering these experiences and histories as sources of knowledge.

The process of studying indigenous knowledges in the context of personal experiences is not an easy one, as it requires, first and foremost a well-established trust relationship between the researcher and the group involved in the research. As Freire and Faundez (1989) suggested, intellectuals (and progressive educators) should "soak themselves in this [indigenous] knowledge, [and] assimilate the feelings, the sensitivity" (p. 46) of epistemologies that move in ways unimagined by most Western academic impulses (Semali, 1999). This entails respecting indigenous people and their knowledge, thus in my own PAR, I felt it was necessary to make myself relevant to the indigenous community that I worked with.

The working procedure for the learning circle was interactive group dialogue, in addition to other learning activities, which comprised of (i) sharing their experiences; (ii) consulting community elders and village (government) authorities; (iii) consulting the Tanzania land policies and laws as related to women's rights to land; and (iv) implementing consensually agreed upon decisions. The same group members met ten times to engage in dialogue sessions. In order to foster dialogue and reciprocity between participants and myself, we relied on a number of traditional indigenous practices; in particular conviviality, mutual interdependence, and networking. These practices worked out positively for us throughout the process of establishing rapport and engaging in dialogue.

A CULTURE OF CONVIVIALITY

Some analysts have shown that indigenous people have strong and ongoing traditions of direct participation within their cultures (Mhina, 2005, 2009; Shizha, 2009; Smith, 1999). This was mirrored in the learning circle that the Haya women participated in. While the women's learning circle provided a forum for me to understand the land tenure situation from women's point of view, it also helped the Haya women who participated in the dialogue to see themselves as knowledgeable people. Each one of the participating Haya women had experienced problems of access and control of land and so had a chance to reflect on their experiential knowledge. They used their personal experiences as a starting point to critically reflect on and raise questions about not having equal rights to land as their brothers and sons. Significantly, this led them to be able to discuss some possible alternatives to the current situation.

Within the process of engaging in dialogue during the women's learning circle, it was through conviviality and reciprocity that a good rapport was established between the Haya women and myself. These two practices also led to a good rapport among the women. When we started our dialogue sessions I shared my own story about a conflict between my aunts and their brother over controlling their father's land. I did this with the intention of making them feel comfortable sharing their own stories, which were related in one way or another to my story. Actually, the kind of questions they asked and the opinions they shared about the story of my aunts told me that they immediately began to engage in discussing something they already knew about. This sort of conviviality at the early stage of the research process worked for both of us—I wanted to start from where they were, which is to say from their experiences.

The flow of communication among the women was possible due to their interpersonal skills that enhanced their interactions. As local women listened to each other's story, a feeling of trust slowly began to develop among them. This feeling of trust and connectedness allowed them to be open and to understand one another. As their stories and experiences converged, individual members moved from isolation to connectedness. They were sympathetic as they sought connections with each other and to negotiate and build a sense of community. More importantly, the Haya women's learning circles captured the essence of mutual interdependence, which was grounded on their social cultural norms.

A Culture of Mutual Interdependence

Indigenous people living in their communities tend not to be individualistic. The women's learning circles followed an indigenous perspective in the way they acted and made decisions. Dei et al. (2000) conceptualize an indigenous perspective as the sum of the experience and knowledge of a given social group, which forms the basis of decision making. This knowledge refers to traditional norms and social values, as well as to mental constructs that guide, organize, and regulate the people's way of living and making sense of their world. Every individual in the learning circle was resourceful in her own way; they had different thoughts and they had different kinds of energies. Group members started the learning process with a fund of knowledge rich in experience and detailed understanding of the problem they encounter as well as detailed firsthand knowledge of their social cultural setting (Heron, 1981). It was important to show the women that I respected their own thinking and intuitions.

As we continued with dialogue in the learning circle, a culture of mutual interdependence was also revealed as we learned from one another. I learned with Haya women how to apply their common partial knowledge to the totality of the women's land situation. As Fear and Edwards (1995) argued, "real participation requires a co-generative dialogue where researcher's knowledge, drawing and abstracting from multiple contexts, is combined with insider's knowledge" (p. 842). Much as the research participants and I exchanged information about women's marginality and how can women change their situation, we were coming from different frames of reference. Group members had insider's knowledge resulting from their experience of the problem encountered while I shared information from my professional experience and other theoretical perspectives.

During these learning circles, I attentively listened to the women and I was open to learn from them. Much as I was a good listener in this learning process, I managed to probe and stimulate group members to think beyond what they already knew. For instance, while reviewing some policies on land use and allocation I asked the women, "to what extent the 1999 Tanzanian Land Act, which emphasizes equal rights of acquisition and control of land for women and men, reflects what is actually practiced here at Maruku village"? Such questions pushed them to discover for themselves what they knew and to confront the discrimination against their rights to agricultural land more critically.

In addition, they took control of what they needed to work with in their own way. Throughout their learning circles local women expressed their indigeneity by externalizing what they had internalized because their decisions on the right way to act in a given circumstance were based on practical knowledge of the social rules of their community. They relied on communal sentiments, which were implicit in their social cultural norms, to attend to the meaning and causes of the discrimination against their right to inherit and control agricultural land.

For instance, instead of waiting to be told what to do, the women established their own support group known as Tweyambe, which means let us work together. The task of the support group was to safeguard women's rights against their marginalization. It was a support group that was established to be used as a forum for women to come together in their struggle of seeking and identifying their own abilities and strengths to change their marginalization with regards to land tenure. Haya women wanted to engage in a continuous analytical process of reflecting on women's situation so that structural impediments could come more clearly into view and also to assert their own independent voice.

To test the working of Tweyambe, women prepared a list of all land disputes that they need to handle. All ten women who participated in Tweyambe learning circle had long-term (ranging from 5 to 10 years) family related land disputes that required them to make several trips back and forth to court houses in Bukoba District, with no sign of a positive resolution to their land problems. Thus, participants of the Tweyambe social group consensually agreed to review each participant's case collectively, starting with one participant who had been evicted from her home immediately after the death of her husband. The widow had lived with her husband for more that forty years and was not allowed to cultivate their farm after the death of her husband. The decision of evicting the victim was based on the rules accorded to customary law (the codified customary practices of landholding) of the Haya people. However, participants of the Tweyambe group were not in support of clan council's decision to evict their fellow woman. When the evicted individual shared her troubles with others, the Tweyambe participants recalled their own experiences, and in the process of remembering and understanding their peer's situation, new meanings emerge. What was more important in the women's learning circles was their ability to network and work collaboratively with others.

INDIGENOUS NETWORKING

On the one hand, Semali & Kincheloe (1999) remind us how knowledge is a crucial element in enabling people to have a say in how they would like to see their world put together and run. While on the other hand, in Western models intellectuals go to the library, in indigenous systems, where knowledge is not stored in archives, people consult elders. Dei et al. (2000) remind us that African indigenousness cultivates respect for the authority of elderly persons for the wisdom of their knowledge of community affairs. In many cases it is the duty of the aged to instruct the young and the duty of the young to respect the knowledge of the elders. In addition, Shizha (2009) has shown that in an African context the ability to use community knowledge produced from local history and memories provides skills critical to survival. It is in the similar vein that members of Tweyambe social group realized that they might not be able to change the order of things from within due to their marginalized status in Haya society. Thus Tweyambe members invited two elders to participate in two of their learning circle sessions. The participating women asked the elders questions for clarification on what seemed not to be working for them based on the experiences of their individual land problems. Consultation in this learning circle needs to be understood in the context of participants sharing experiential knowledge and learning from each other's knowledge and skills and not women seeking answers from elders. Below are some of the questions women asked their elders and some of the issues discussed during the dialogue within the learning circle (P4 and P10 stand for the women who participated in the dialogue).

P4: If customary law allows women to inherit land how come most women, particularly widows and single unmarried women, are still denied their rights to land? . . . There are some incidents where the husband dies and when children grow up they decide to throw their mother out of the home. What are your comments on this issue and what is a mother supposed to do?

P4: In most cases the main heir is the eldest son and it so happens that after he marries, his wife tends to harass the mother in law. What was the situation during our parents' generation?

P10: I have seen cases where widows are very problematic and they engage in endless fights with their daughter-in-laws. What would be the best way to solve interfamilial conflicts?

Generally village elders were attentive and sympathetic to women's concerns taking into consideration that these concerns were based on the knowledge of what the women experienced in terms of discriminating land rights, and the fact that most of the changes introduced in the land management and land dispute were more disadvantageous to women—who are either their daughters, daughters-in-law, or granddaughters. Part of the elders' contribution in terms of exchanged information was the description of the openness and flexibility of the inheritance system as opposed to rigid legalized Western models of land holding. For instance, they gave an example of how *matanga* meetings were successfully conducted in the past. In Haya society, when a person dies, a meeting called *matanga* is held three days after the death. The *matanga* is part of the general administration by the clan council at which a disposition of the properties, claims, and debts of the deceased is made. Thus, the final decision on who should inherit a particular *kibanja* (the permanent homestead) rests with the patrilineage or the clan council. During the discussions the elders emphasized how in the old day the decision of the clan was respected and followed and how the traditional authority was devalued and despised as people learned how to manipulate the new formalized land tenure systems. Elders also believed that the old dispute resolution mechanism seemed to have worked because it was a framework of organizations, relationships, cultural ideas, guidelines, and prescribed rules of preference together with conceptions of morality. These were all intertwined in the web of ordinary activities. The elders also emphasized that it was highly shameful for the people and the entire community to not have their lineage authority accepted and utilized appropriately. Elders also clarified the inefficiency of the legal system and how men use their financial power and knowledge to manipulate the system.

It was from the information contributed by the elders that Tweyambe participants brought up and discussed with elders the idea of reinstating the traditional mechanisms of handling family issues, including land disputes, which was viewed as better and more efficient than the court system. Based on their experiences, Tweyambe participants considered the court system to be humiliating. One participant shared her experience of being raped by and two others experienced physical assaults from their opponents in the land disputes they were involved in. In their discussions with elders, Tweyambe participants seemed to be comparing how things operated in the past and what was currently happening in their situation, and based on their brutal experience with the imposed court system, they were inclined to perceive the system of the past as much more effective and fair.

To me this act of engaging in dialogue with the elders was a meaningful recovery of history, which came about through the collective remembering of those elements of the past that proved useful, thus supporting traditional institutions. What was important at the time of discussion was that the elders were very supportive of the women and agreed to work collaborative with them to have one voice toward making a change.

Tweyambe group members' collaboration did not end with the consultation with the elders. They formed social networks with their acquaintances and fellow women who had not participated in the learning circle. They invited women from other *vitongoji* (neighboring villages) to attend what they called a Tweyambe day, whereby participants of the learning circle communicated to others about Tweyambe and how women can benefit through Tweyambe. I interpreted this act of dissemination of knowledge and mobilization of other women as a spontaneous spread of a movement from village to village. This signified a horizontal coalition and a collective initiative of bringing about change for women. Furthermore, Tweyambe members created contacts with women lawyers and other related institutions at the regional headquarters to exchange information and become more informed about their rights. The motive was to build alliances and networks that would allow Haya women to put collective pressure on unresponsive local leaders and unfamiliar imposed Western institutions.

Dei et al. (2000) has shown how traditional social groupings provide social comfort, identity, and a sense of belonging to a community, particularly in times of stress and hardships. In similar vein, Tweyambe participants sought connections with each other to build a sense of community. The implementation of collaborative ventures was possible because Tweyambe group members did not work as individuals; rather they worked as a team.

Throughout the learning process, group members demonstrated empathy and sympathy and it was the compassionate feelings among women that enhanced their collective strength and confidence, which Kabeer (1999) calls power within. The more they saw themselves capable of producing and defining their own reality, the more activated they became to change it. The next section will briefly discuss the outcome of the learning process and the collaborative effort of Haya women and the rest of the community.

HOW DID KNOWLEDGE ACTIVATION CHANGE THE EXISTING SITUATION?

SOCIAL TRANSFORMATION

The objective of PAR has been well articulated by various authors as leading to social transformation and change (Hall, 1993; Mhina, 2005, 2009).

Social transformation, for the Haya women, emerged as a process and a product of the dialogues and interactions with one another as well as with myself. The object was understanding their problematized situation and interacting with different actors, including the village elders, and government representatives at the village level, each bringing diverse and partial knowledge to the dialogue.

By taking part in the PAR learning process, Tweyambe participants revealed their agency and what I have elsewhere called inarticulate intelligence (Mhina, 2009), which had been encapsulated in their uneasiness and misery of their marginalized condition. It can be said that these Haya women took initiative to voice out their concerns about their marginal position within the structure of indigenous land tenure system (Mhina, 2005) and by applying their indigenous knowledge and incorporating the local history and elders' experiences and wisdom they challenged drawbacks of both indigenous and the court system. They have embraced new innovations, in particular implementing women's rights to inherit land, yet they have done so in ways that supported and maintained their tradition, such as altering the way the land disputes were handled in their village without interfering with the village social organization. For instance, Tweyambe members, as a group, convinced the village government to revoke the decision of the clan council and to review the case of the evicted woman from her home after her husband's death. These are huge accomplishments given that previously, all Tweyambe members were trapped in their powerlessness and isolation. However, after engaging in the learning circles, they recognized the significance of working collectively. It was through the catalytic intervention that Haya women were able to begin work to change their marginalized situation. The actions implemented by women did not interfere with the rules of the lineage system nor did they undermine the basic principles of social organization and cohesion. As Freire and Faundez (1989) remind us, we need to accept that knowledge possessed by the people represents a rich social resource for any action to change the society.

LOCAL AUTONOMY

This case study also illustrated the ability of Haya women to manage their own affairs. As they realized they could do this, they also demonstrated a growing interest in doing so. Haya women were able to utilize their indigenous human and social resources (Kabeer, 1999), in particular, the hidden, practical knowledge, which was firmly grounded on the realities of their social and cultural lives, to generate knowledge that was useful to address the land issue they were facing. It is important to note that the change that they effected in their situation was not done to them and the knowledge generated by Tweyambe participants was not given to them. Instead it was a consequence of their effort to engage in a collective struggle. This case echoes Fals-Borda and Rahman's

(1991) argument that people cannot be liberated by a consciousness and knowledge other than their own. It also illustrates how the unrecognized and unacknowledged talents and skills that women make use of in their daily lives can be lifted to the conscious level not only for the women themselves, but for the community as a whole. This process of bringing knowledge forward can also be viewed as part of a broader process of democratization and building local-level political and decision-making capacity.

CONCLUSION

This chapter is an account of indigenous women's ability to address discrimination issues against their rights to land. Based on the catalytic intervention of PAR and women's willingness to reflect and act, they discovered the need to denounce their marginality by thoroughly reevaluating and reinterpreting their own experiences. My role as researcher was to initiate the PAR process and to inspire women to reflect on the marginal land rights that affected their lives and to act upon their understanding and decisions to change their situation.

In this chapter, Haya women are credited with the capacity of using their indigenous knowledge to devise ways of tackling the problematic situation they were facing. They experienced their capability and power to produce knowledge autonomously. The knowledge generated by Haya women was based on their customs and traditions and thus was more appropriate and useful in their situation.

Women had the opportunity to recover their wisdom, which they managed to turn into a potent force for emancipating the rest of the community members, thus helping to create a fuller life and a more just society. To some extent this revival of their wisdom and confidence of implementing actions for themselves reduced their dependence on external experts. Therefore, through this process, not only the women, but also the society as a whole became more self-reliant. I argue that if academic and popular knowledge are combined, this may result in useful knowledge of an emancipatory nature, which can be used as a rich resource for addressing issues and to bring about social change.

What my research has shown is that local women's knowledge is a tremendous social resource, which no society can afford to undervalue any longer. Furthermore, an emancipatory approach to development should be able to build on the ability of local people's ability to generate and apply their indigenous knowledges, which are grounded in their culture and social histories. Learning from indigenous people and instigating change from the grassroots is a meaningful way of enhancing interdependence between professional and indigenous people. The question is whether we professionals are prepared to learn from and to tap into the creativity and resourcefulness of diverse local groups.

REFERENCES

Bruce, J. W. (1993). Do indigenous tenure systems constrain agricultural development? In T. J. Bassett, & D. E. Crummey (eds.), *Land in African agrarian systems*. Madison, WI: University of Wisconsin Press.

Dei, G. J. S, B. Hall, & D. G. Rosenberg (2000). *Indigenous knowledge in global contexts: Multiple readings of our world*. Toronto, ON: University of Toronto Press.

Fals-Borda, O., & M. Rahman (eds.) (1991). *Action and knowledge: Breaking the monopoly with participatory action research*. New York: Apex Press.

Fear, K., & P. Edwards (1995). Building a democratic learning community within a PDS. *Teaching Education, 7*(2): 13–24.

Freire, P. (1973). *Education for critical consciousness*. New York: Seabury Press.

Freire, P., & A. Faundez (1989). *Learning to question: A pedagogy of liberation*. New York: Continuum.

Hall, B. (1993). Introduction. In P. Park, M. Brydon-Miller, B. Hall, & T. Jackson (eds.), *Voice of change* (pp. xiii–xxii). Westport: Bergin and Garvey Publishers.

Heron, J. (1981). Philosophical basis for a new paradigm. In P. Reason & J. Rowan (eds.), *Human inquiry: A source book of a new paradigm research* (pp. 19–35). Chichester: John Wailey and Sons.

Kabeer, N. (1999). *The conditions and consequences of choice: Reflections on the measurement of women's empowerment*. Geneva: United Nations Research Institute for Social Development.

Kapoor, D., & S. Jordan (2009). Introduction: International perspectives on education, PAR, and social change. In D. Kapoor, & S. Jordan (eds.), *Education, participatory action research and social change: International perspectives* (pp. 1–11). New York: Palgrave Macmillan.

Lugalla, J. L. P., & M. A. C. Emmelin (1999). The social and cultural contexts of HIV/AIDS transmission in the Kagera Region, Tanzania. *Journal of Asian & African Studies, 34*(4), 377–395.

Mamdani, M. (1996). *Citizen and subject: Contemporary Africa and the legacy of late colonialism*. New Jersey: Princeton University Press.

Mhina, C. H. (2005). *Social learning for women's empowerment in rural Tanzania*. Unpublished doctoral dissertation, University of Alberta, Edmonton, Canada.

——— (2009). "Research and Agency": The case of rural women and land tenure in Tanzania. In D. Kapoor & S. Jordan (eds.), *Education, participatory action research and social change: International perspectives* (pp. 155–168). New York: Palgrave Macmillan.

Mhina, C. H., & A. A. Abdi (2009). Mwalimu's mission: Julius Nyerere as adult educator and philosopher of community development. In D. Kapoor & A. A. Abdi (eds.), *Global perspectives on adult education* (pp. 53–69). New York: Palgrave Macmillan.

Migot-Adholla, S. E., & M. E. John (1994). Agricultural production and women's rights to land in sub-Saharan Africa. In Sylvia Huntley (ed.), *Development Anthropology Network: Bulletin of the Institute for Development Anthropology, 12*(1&2), 14–29.

Mikell, G. (1997). Introduction. In G. Mikell (ed.), *African feminism* (pp. 1–50). Philadelphia: University of Pennsylvania Press.

Peters, P. E. (2004). Inequality and social conflict over land in Africa. *Journal of Agrarian Change, 4*(3), 269–314.

Purcell, T. W. (1998). Indigenous knowledge and applied anthropology: Questions of definition and direction. *Human Organization, 57*(3), 258–272.

Semali, L. (1999). Community as a classroom: Dilemmas of valuing African indigenous literacy in education. *International Review of Education, 45*(3–4), 305–319.

Semali, L., & J. L. Kincheloe (1999). Introduction: What is indigenous knowledge and why should we study it? In L. Semali & J. L. Kincheloe (eds.), *What is Indigenous Knowledge?: Voices from the Academy* (pp. 3–57). New York: Falmer Press.

Shizha, E. (2009). Critical analysis of problems encountered in incorporating indigenous knowledge in science teaching by primary school teachers in Zimbabwe. *Alberta Journal of Educational Research, 53*(3), 302–319.

Smith, L. T. (1999). Decolonizing methodologies: Research and indigenous peoples London: Zed Books Ltd.

Toulmin, C., & J. Quan (eds.) (2000). *Evolving land rights, policy and tenure in Africa.* London: DFID/IIED/NRI.

CHAPTER 14

THE INDIGENOUS KNOWLEDGE SYSTEM OF FEMALE PASTORAL FULANI OF NORTHERN NIGERIA

LANTANA M. USMAN

INTRODUCTION

Recent debates on the role and contribution of African Indigenous Knowledge Systems (IKS) have expanded the scope of African studies and education scholarships. African pastoral communities, which consist of indigenous people, are a few of the continent's population that retain and promote IKS, not only as a means of cultural sustainability, but for survival and general livelihood. This chapter examines the feminization of indigenous knowledge systems (IKS) among the pastoral Fulani of Nigeria, as well as across West Africa. Through literature reviews and oral narratives, the chapter elucidates women and girls' cognitive, aesthetic, spiritual, and moral educational components of cultural literacy. These are not only the norms and values of the Fulani females, but considered the cultural essentials that define females' social agency and coexistence. In addition, the chapter recapitulates the purpose, nature, characteristics, and processes of these knowledge systems in order to connect the learning outcomes to the females and the ethnic community in general.

A world classification of indigenous people included pastoral Fulani, the Wodaabe and the Bororo'en of West Africa (International Work Group for Indigenous Affairs [IWGIA], 1998/1999). The nomads' way of life is centered and recycled around IKS of cultural, economic, and social life. IKS are essential to nomads' practical realities and existence, thus are part of a lifelong learning. As noted by Nafukho, Amutabi, and Otunga (2005), most African IKS are the basis of a cultural literacy that is in harmony with social, political, and economic sustainability. Often, learning experiences are shaped by the ethnic group's philosophy of life (Chukwu, 2002). The Fulani pastoralists' philosophy of life is connected to religious idealism of nature and purpose, as well as the existence of humans' harmony with the creator and nature (environment); it is synonymous with Aristotle's construct realism (Gutek, 2005). The ramification of IKS for the nomadic Fulani is a continuous learning process and outcomes that are not only gender specific, but also gender sensitive. From the tribe's oral historical origin of the *Koforundu* chronicle, it is stated that the development of the Fulfulde language of the ethnic group originated from the first Fulani woman and her children, thereby placing women as first teachers and custodians of maternal pedagogies, which makes Fulani women a significant group in their community (Junaid, 1987; Usman, 2005).

In light of the above synopsis, the chapter analyzes major IKS epistemologies (cognitive knowledge), axiology (value system), aesthetics (goods, beauty, and arts) and the idealism-realism construct (spirituality and the environment) practices of the pastoral Fulani women and girls. The feminization of the knowledge system is significant to the social and economic sustainability of the women's lives and the promotion of their identities. The account explores literature and oral historical discourses of the women and girls' IKS and their contributions to the scholarship of world IKS.

THE SOCIAL SPACE OF PASTORAL FULANI WOMEN

Fulani nomadic females (women or *rewbe* and girls or *surbaaBe*) of West Africa are a significant indigenous population of sub-Saharan Africa (IWGIA, 1998/9; Stenning, 1994). In northern Nigeria, they constitute half of the population of 9 million upland nomads (Usman, 2005), and are spread across Lake Chad, the prairies of Adamawa Highlands of Nigeria, and the Republic of Niger and the Cameroon (Ibrahim, 1966; Stenning, 1994).

The women or *rewbe* (in Fulfulde the language of Fulani) have distinct social identities and statuses are not only ascribed from birth, but they are also labeled within the patriarchal institution of the ethnic group. Females are placed in the lower socioeconomic class and are considered as secondary providers of the nuclear and extended family systems. Their sustainable informal economies are centered on dairy management and marketing, which provide females with a source of income and livelihood to support themselves as well as the family as secondary providers and nurturers (Dupire, 1971; Usman, 2005). Like other pastoral women of sub-Saharan Africa, Fulani pastoral females are more respected and their social statuses greatly recognized especially when married and hence young girls are expected to aspire to become wives (de St. Croix, 1972; Dupire, 1971; Kipuri, 1991). Furthermore, female's marital social identities are expected cultural norms and values that are displayed through behavior and appearances. All females engage in various types of culture-based aesthetic body modifications such as facial tattoos, hairstyles, and clothing that distinguish their age, stage of marriage, or right of passage (de St. Croix, 1972; Dupire, 1971). For example, little girls wear *durol cakaol* (a special braided hairstyle) while married women and girl-child wives wear a hairstyle called *durol bedyeli pu DaaDo* (married women's hairstyle) (Junaid, 1987).

Female body modifications aim at maintaining and enhancing their natural beauty. As noted by de Villiers and Hirtle (1997), "they have their hair plaited, a lipstick of *kohl* blackens their lips, and yellow paste made from a friable stone called *polla* is spread all over their face" (p. 285). The female practices represent major cultural values that define their femininity and sexuality and that are also greatly appreciated by the opposite gender, who believe that these practices are part of the cultural norms, mores, and values that are central to the tribe's aesthetic value knowledge and skills. These aesthetic value systems are transmitted from older women to younger women through imitation, observation, and role enactment as methods of teaching and learning (de St. Croix, 1972; Usman, 2005).

RAMIFICATIONS AND LEARNING OUTCOME OF THE FULANI WOMEN'S IKS

Prior to the colonial invasion of Africa, the people had an educational system that was functional and based on lifelong learning in preparation for adulthood and the social responsibilities accorded to that stage of life. The aim of learning was and still is for the induction of members of the ethnic groups into their societies and to prepare them for adulthood (Fafunwa, 1974). Fafunwa added that "African education emphasized social responsibility, job orientation, political participation and spiritual values" (p. 13). African IKS are both cognitive and practical, and are based on cardinal goals and principles of: developing a child's latent physical skills and character; inculcating respect for elders and those in positions of authority; developing intellectual skills; acquiring specific vocational training and instilling positive attitudes toward honest labor; fostering of belonging and active participation in family and community affairs; and understanding, appreciating, and promoting cultural heritage of the community at large (Fafunwa, 1974). Dei (1996) explains,

> Indigenousness is knowledge consciousness arising locally and in association with long-term occupancy of a place. Indigenousness refers to the traditional norms and social values, as well as mental constructs which guides, organize and regulate African ways of living and making sense of their world.

Indigenous knowledge differs from conventional knowledge because of an absence of colonial and imperial imposition. The notion of indigenousness highlights the power relations and dynamics embedded in the production, interrogation, validation and dissemination of global knowledge about international development. It also recognizes the multiple and collective origins of knowledge and affirms that the interpretation or analysis of social reality is subject to different and sometimes oppositional perspectives. (p. 1)

IKS according to Nafuko, Amutabi, and Otunga (2005) are "knowledge and skills that encompass the entire survival and coping mechanism of a [tribe] community, and are acquired outside the formal educational system" (p. 18). In addition, Coombs and Ahmed (1974) refer to IKS as:

The life long processes by which every person acquires and accumulates knowledge, skills, attitudes, and insights from daily experiences and exposure to the environment at home, at work, at play, from the examples and attitudes of family and friends . . . Generally, it is unorganized, unsystematic and even unintentional at times, yet accounts for the great bulk of any person's total lifetime learning including that of even highly "schooled" persons. (p. 8)

African IKS are grounded in each ethnic group's philosophy and wisdom, both which are passed from one generation to the other (Chukwu, 2002; Fafunwa, 1967, 1974). Specific to pastoral Fulani women the teaching and learning of IKS begins from childhood and continues to old age. It involves learning the culture and traditions of the past, which are cognitive and practical perspectives that are participatory and active, and based on imitation, observation, listening, doing, recitation, and demonstration. It is a learning engagement that is also part of socialization, in which gender, age group, and rite of passage are marked as learning processes and outcomes. The primary agents of "deschooling" of the nomadic female Fulani are the family (extended and nuclear), peer groups, and the mosque. The praxis of teaching and learning is directed by maternal pedagogies, as mothers, elderly women, and other significant females of the family and community serve as primary and secondary instructors.

COGNITIVE KNOWLEDGE SYSTEMS

The general epistemologies of the girls and women's cognitive knowledge are broad and interdisciplinary. The learning processes are directed by rote learning or memorization, observation and imitation of the latent and implicit curriculum, as well as applying and practicing skills (Fafunwa, 1974). The modeling of learning is based on the three Ls: Look, Listen, and Learn (Fafunwa, 1974; Usman, 2005). For example, in the process of teaching young girls dairy management, older women demonstrate the milking process, churning of butter, and fermentation of milk using local utensils. The young ones observe and listen and are eventually allowed to practice what they have observed.

Curriculum content include (but are not limited to) cognitive learning of proverbs, which are short, repeated, witty statements or sets of statements of wisdom, truth, and experience which are used to further a social end (Yusuf, 1998). Proverbs are idiomatic expressions, popular tribal sayings, and songs that require memorization and meta-cognition (Ezeomah, 1983). As observed by Ojoade (1982), parents teach proverbs to epitomize a moral lesson they wish to impart on their children. The practice is general to most African rural ethnic groups. As noted by Dei (2000) "elders taught oral literature, fables, folktales, legends, myths, proverbs and story telling. African youths receive socialization and education" (p. 13).

Ethnomathematics is one example of a learning process that is based on maternal pedagogies. Mothers/significant female adults teach ethnomathematics to nomadic Fulani girls through the mental exercise of counting, and the action of knotting rope(s) tied by the bedside of the child or the mother/ significant female adult. Dupire (1971) added that "mental education is given to her [girl] by her mother who answers all her questions, gives her practical training in the use of customary equipment and teaches her how to count by means notches cut on the bed pole" (p. 55). The purpose of learning ethnomathematics is to enable girls to acquire basic orthodox numeracy, as well as learn and keep track of days and dates on the calendar (both modern and Islamic types) so as to know when market days are so they can sell their dairy products of milk, cheese, and butter. Significantly, counting days through knotting of ropes provides information that enables the female adults and adolescents to monitor their reproductive cycle; that is, monthly menstrual cycles as well as expected date of delivery for pregnant mothers.

This traditional learning style is used by primary school attending Fulani girls in subjects like Arithmetic. Commercial teaching and learning manipulative aids are expensive and not available in most rural primary schools in northern Nigeria, hence, students and teachers adapt the indigenous process to accommodate the deficiency of learning resources. The major advantage of incorporating indigenous learning techniques is that they accommodate IKS and are centered on continuity of learning style, flexibility of learning materials, and hands on learning for a more pragmatic and constructivist educational philosophy of learning (Ryan & Cooper, 2007).

SOCIAL MEDICINES AND SPIRITUAL KNOWLEDGE SYSTEM

The concepts and practices of orthodox medicine and trado-pharmacology represent a lifelong learning that promotes sustainability of most if not all African ethnic groups (Oliver-Bever, 1986; Sofowora, 1993; Togbega, 2006).

Pastoral Fulani female practitioners of the system are accorded special status, social prestige, and autonomy in their communities (Worley, 1991), especially with those whose family lineage are ascribed with the practices (Fafunwa, 1974; Sofowora, 1993). With the female Fulani pastoral women and girls, the nature and characteristics of their migrant life makes it mandatory for them to be prepared for health challenges that include illnesses such as malaria, common colds, rheumatism, and diarrhea, which are the most common in tropical environments. Their caring practices extend into their roles as family nurses who often administer medications of herbs, oils, powders, and allied traditional medicines to themselves and members of their families. Their knowledge and practices of traditional medicine are demonstrated in their ability to select, process, and store tree shrubs, leaves, roots, barks, traditional oils and fats, as well as powders of animals and reptiles, and liquids for curative purposes and prevention. For example, the women use leaves of *neem or azadirachta* [referred in Hausa lingua franca of northern Nigeria Hausa as *dogon yaro*] and *carica-papaya* (paw-paw leaves), and guava leaves (*p. guajava*) in treating malaria, diarrhea, and stomachache for themselves, their children, and other members of their households (Ekanem, 1978).

The women and girls' role in cooking has empowered them with knowledge of spices that are not only used as food, but also as curative medicines. It is a cultural value and normative expectation and practice for female pastoral women to possess spices such as black pepper (*piper guineense/piperaceae*), also referred in Hausa language as *masoro/kaninfari,* and ginger (*zingeber offcinale/zingiberacene*), also referred to as *chitta* in Hausa language. These basic and necessary spices are stored in the calabashes of all females, especially during migration. The women use these spices for the treatment of common colds/flu, joint pains, and cough therapy for themselves and members of the family. The nature of their constant migration exposes them to harsh weather that triggers illnesses such as cold/flu, which affect mostly children who are constantly in the company of the women. It is the responsibility of Fulani women to provide them with immediate therapy or medicines using such spices and herbs.

Additionally, female Fulani traditional midwives or *angwanzoma* in Hausa language, use and administer selected bitter rich herbs mixed together and referred to in Hausa as *madachi* [bitter concoction] as dietary supplements for pregnant women, and lactating mothers. Also, the administration of selected herbs assists pregnant women's muscles and nerves to relax during labor contractions. It is also believed that the herbs act as cleansing products of toxins in the systems of pregnant women, thereby ensuring a healthy baby at and after birth, as the mother continues to consume the herb solutions for a year. Also, post delivery care of new mothers involves a traditionally administered hot bath or (*wankan jego*) as body therapy (Kisekka, 1992). The administration of body massage and therapy are performed by experienced women with some basic knowledge in traditional midwifery. The process involves the use of hot water soaked in a solution of specific therapeutic tree shrubs and leaves of and *carica-papaya* (paw-paw), and *azadirachta indica or* neems also referred as *dogon yaro* in the Hausa lingua franca, used by the Fulanis. The bath process lasts forty days and is performed on the new mother twice a day. Often, the skills and knowledge of preparing and giving the hot bath are passed from one generation of women to the next. This is a means of preserving the knowledge and skills.

Additionally, the hot baths are believed to be analgesic, to restore energy, and to rejuvenate blood circulation of the mother's body, after the long period of labor. Lactating mothers also use a lot of medicinal spices in their puddings and soups to provide and restore energy after birth, prevent breast milk from clotting into painful lumps (the spices sensitize the flow of breast milk), as well as protect the breast fed baby from catching a cold. The traditional reproductive knowledge systems of the women have sustained mother and child health care for years, and are still practiced today, despite being undervalued and despised by modern medicine (Kisekka, 1992).

Fulani and Hausa rural women and girls are knowledgeable about practices of traditional spiritual education, with "spirituality knowledge connected to humility, healing and empowerment and value of life" (Dei, 2002, p. 4). The women's orthodox spiritual knowledge systems are based on the *Bori* occult system that involves the control of forces and the performance of adoricism in dance and music in which the spirits are controlled to heal illnesses (Palmer, 1914). Women priestesses serve as mediums to those seeking assistance with healing illnesses and diseases, emotional therapy, and spiritual assistance. During worship, the women are clothed in special black traditional woven cloth and dance to the rhythms of local guitar known as *garaya* by raising themselves by jumping into the air as high as six feet above the ground and falling into trance. The ritual symbolizes a period of divine contact and receiving of revelation which is followed by incantations and divine messages of a spiritual nature, which often lead to laying of hands on the body or touching the sick in a process of healing (Frank, 1975; Umar, 1999). Even though *Bori* priestesses are Muslims, they achieve new status, self-empowerment, and social roles as divine healers, which earn them a lot of respect in the community (Lewis & al-Safi, 1991). The older female priestesses pass on the knowledge and skills to younger girls through *Girkaa* initiation (Veit & Habou, 1989). *Girkaa*, the process of inducting younger females into adulthood, has not only increased the knowledge continuity of female participation, but the domination of the female gender in the *Bori* system (Umar, 1999).

Female nomadic Fulani women wear talismans and amulets that contain verses of the Quran. The ornaments are supposed to protect them against sickness, premature birth of babies, bad luck in trading and marriage, as well as general family protection (de St. Croix, 1972). The use of symbolic traditional social medicine is accompanied with knowledge responsibilities.

Women are expected, when the need arises, to memorize and spontaneously recall verses of the Quran. Such practices are based on cognitive learning that requires personal commitment. The memorization task of the talisman or shielded verses of the Quran is not only a mental activity that is personal, but it is also an inherited knowledge practice that is passed from mother to daughter, amplifying and symbolizing knowledge continuities.

In sum, women and girls' practices of spiritual education are a knowledge base, a function of healing illnesses, and an emotional therapy for members of the family and ethnic community in general. As common to most African communities, "[s]spiritual education embraces humility, respect, compassion and gentleness that strengthen the self and the collective human spirit of the learner" (Dei, 2002, p. 6).

ENVIRONMENTAL EDUCATION

The migrant life pattern of the Fulani enforces them to be conscious and knowledgeable about their social and physical environment, so as to determine their choice of seasonal settlement, and with special attention to their cattle (Stenning, 1994; 1957). Even though emigration and settlement decisions are the jurisdiction of men, as companions and spouses, women are often informed of the nature and characteristics of the land topography, rivers, and climatic changes, especially with relation to proximity to urban areas, as the women and girls are expected to visit these places to market their dairy products.

Additionally, nomadic Fulani girls are taught about land, rivers, and types of grass and shrubs that the cattle can consume during herding. The girls are taught the position of water holes and how to estimate their capacity, how to identify grazing lands and their composition of nitro-rich soils for curative purposes, and how to read the clouds for rainfall and winds. This indigenous knowledge is very valuable for smooth adaptation to the new environment and pertinent to continuity in community economic activities and social life. The purpose of learning and becoming knowledgeable about weather information and interpretation is to help the girls prepare and move to market centers to sell their dairy products. Girls and women gather wood and shrubs for cooking fuel and in the course of fuel gathering they learn to distinguish different types of corn stalks for fuel and cattle consumption. Mothers and elderly women teach girls the types of vegetation and trees so that this knowledge is carried down from one female generation to the next. Due to their emigration pattern, women explore the environment and identify and collect wild fruits and leafy vegetables as sources of food. While food is vital for the sustenance of the group, women also collect special twigs for rope making (de St Croix, 1972; Dupire, 1971).

The pastoral Fulani consider cows as special gifts from the creator. Both women and men respect cattle. The tribe believes in the supernatural powers of cattle in sustaining their livelihood and guiding them in deciding and selecting suitable and safe environments for settlement and grazing settlement (Stenning, 1994). As a result of their dependence on cows for sustainability, women and girls consider them special, and often refer to a cow as "she," as a way of bonding socially with the animals. Due to the women's continual access to cattle milk, which facilitates their dairy processing of sour milk, curds, and butter (Shiva, 1988), pastoral Fulani women consider the calves to be part of the family and community and nurture calves as they nurture their children (de St. Croix, 1972). Another reason for the women's commitment to nurturing the calves is to ensure their good health and growth for future production of milk, for domestic consumption (as a stable food) and for processing and distribution in urban markets as a primary source of income for the women and girls (VerEecke, 1991).

Migration is not only an economic necessity for women in Fulani families, but also for the whole Fulani nomadic group. Relocation itself is also means of educating and exposing women to the world of nature as well as learning "self perseverance in the face of all hardships" (Junaid, 1987, p. 28), which is the core of the Fulani moral code of *pulaaku*. Women learn to read their natural environment and learn to make sense of it from the knowledge accumulated and transmitted from one generation to another. The natural environment is a source of life for the whole group and their cattle. Environmental knowledge embedded in their indigenous epistemology is a source of livelihood and social and economic continuity.

KNOWLEDGE OF ARTS AND CRAFT

The aesthetic skills and expression of beauty and arts and crafts of the nomadic Fulani women are a cultural feminine value. Pastoral Fulani women value their appearance in terms of beauty, which often is enhanced through traditional cosmetics, and creative body modification (de Villiers & Hirtle, 1997). Furthermore, the women and girls' symbolic art creations are demonstrated on their calabashes, the main domestic and commercial utensil. The women create decorative patterns, using cold local tools, such as kitchen knives and rock sharp objects, and white native chalk paste made from dry milk. Their artistic designs are combinations of various abstract motifs such as circles, squares, triangles, and lines, and animal figures such as lizards and snakes that are symbolic of the tribe's philosophy of life (Beckwith & Fisher, 1999; Mbahu, 1999). In addition, Jest (1956 cited in Mbahu, 1999) observed that:

> The geometrical motifs are deeply appreciated by the Fulbe. The circles, triangles and jagged lines are reproduced on the walls of their huts, materials and poetry designs. This geometrical pattern is certainly an expression of the deep preoccupation of their intellect. (p. 45)

Further images displayed in the paintings and designs include plants with leaves, human figures, camels, horse riders, and women pounding food in a traditional mortar. The display of the women and girls' imaginary creativity further highlights their knowledge construction in aesthetics and axiological branches of philosophy of education. The generated patterns and designs reflect a social realism with respect to nature and the community, which is central to most IKS in African ethnic groups.

MORAL EDUCATION

For female pastoral Fulani, moral education is related to character education and is guided by behavior conduct, etiquette, and general mannerisms that define and construct their social identities. The ethnic moral code of *pulaaku* is courage, strength, humility, perseverance, good character, bravery, and self-empowerment by all Fulani women and girls. *Pulaaku* practices are demonstrated through positive behavioral conduct and effective communication in private and public spaces, such as being good listeners and not making direct eye contact with elders during conversation to show respect. The Fulani believe *pulaaku* calms their minds and inner spirits, thereby allowing them to have control over the negative social vices of anger and hatred. Since pastoral Fulani take *pulaaku* very seriously, it is generally considered a social control mechanism.

Women or *rewbe* are expected to be of good character and are expected to display confidence, reservation, and shyness during public conversation (Dupire, 1971). Young girls or *surbaaBe* between the ages of four to five years are taught essential rules of the sociomoral code *mbo Dangaku*, which demands them to conduct and communicate with the opposite gender appropriately to maintain self dignity and family honor (Junaid, 1987). Additionally, females learn moral education through proverbs. Often, nomadic Fulani women and girls of good character command respect from their relations and in-laws. In the case of young unmarried girls their good morals are rewarded during the selection of marriage partners when Fulani bachelors compete in a game dance of *sharo* (a male stick whipping dance) meant to test manhood as well as to win the choice of a bride. In sum, moral education taught through *pulaaku* is what distinguishes the pastoral Fulani's social identity from other pastoralists such as the Tuaareg in West Africa.

DOMESTIC TRAINING

Traditional practices of education are an integral aspect of everyday life and serve primarily for the transmission of traditional roles and skills. Mothers and older women provide training of home economics or domestic science to young girls during the first years. The girls acquire survival domestic skills as part of their daily family chores as they learn and demonstrate how to fetch water and firewood, prepare food, milk the cows, pound and grind grains, churn butter, and clean and take care of younger brothers and sisters. In families, young girls are expected to spend most of their time at home doing domestic chores and may not leave the household unless authorized by parents to do so. As mothers make regular market visits to sell products and are absent from home, girls cover their mother's domestic duties in addition to their own duties (Junaid, 1987).

On the other hand, older girls learn the skills of family maintenance and nurturing as they have the role as substitute caretakers of the younger children and infants during the absence of their mothers. If a girl were to leave the infants for any reason she would be punished by her mother (Junaid, 1987). A number of the Fulani childcare techniques are used in their married homes when they become mothers and wives. These tasks are considered essential for the functioning of female social life among the nomadic Fulani as well as preparing the young girls for motherhood. The practice is handed down from generation to generation, a demonstration of essential knowledge and skill continuities, which are central to most African ethnic tribes.

CONCLUSION

Traditional African knowledge systems are broad and interdisciplinary. Their practices are affected by differences in geographies but are in harmony with the need for sustainability, continuity, and livelihood. The chapter recapitulated the feminization of IKS with the pastoral Fulani women. An overview of the epistemologies, axiology, aesthetics, and idealistic religious philosophies that are connected to spirituality and healing, moral education, and environmental education were discussed in relation to the purposes, practices, and development of the women and girls. The continuity of these knowledge systems and practices of the women and girls indicates continuity of the cultural tenets and heritage of the tribe. It demonstrates that the impact of modernization with the nomadic women has boundaries, which have allowed them sustain their IKS. The discussion also provides a voice for the women, as well as a source of information that may be considered for further research inquiries that may be beneficial in the scholarship of deschooling and development in Africa. It is the resolve of this discussion that more attention should be provided to women and girls' practices of IKS in Africa, as little is documented or studied about them, when compared to the opposite gender.

REFERENCES

Beckwith, C., & A. Fisher (1999). *African ceremonies* (Vol. 1). New York: Abrams Publishers.

Chukwu, C. N. (2002). *Introduction to philosophy in an African perspective.* Eldoret: Zapf Chancery Research Consultants and Publishers.

Coombs, P. H., & M. Ahmed (1974). *Attacking rural poverty. How nonformal education can help.* Baltimore: John Hopkins University Press.

de St. Croix, F. W. (1972). *The Fulani of northern Nigeria.* Farnborough: Gregg International Publishers.

de Villiers, M., & S. Hirtle (1997). *Into Africa. A journey through the ancient empires.* Toronto: Key Porter Books.

Dei, S. J. G. (1996). *African development. The relevance and implications of indigenousness.* Paper presented at Learned Societies meeting of the Canadian Association for the Study of International Development (CASID), May 31–June 2, 1996. Brock University, St. Catherines, Ontario.

———. (2000). Rethinking the role of Indigenous Knowledge in the academy. *International Journal of Inclusive Education, 4*(2), 111–132.

———. (2002). *Spiritual knowing and transformation learning. New Approaches to Lifelong Learning (NALL).* Working Paper Number 59. OISE/University of Toronto. Retrieved from, www.nall.ca.

Dupire, M. (1971). The position of women in a pastoral society (The Fulani WoDaaBe, nomads of the Niger). In D. Paulme (ed.), *Women of tropical Africa* (pp. 42–53). Evanston: Northwestern University Press, Berkeley.

Ekanem, O. J. (1978). Has Azadirachta indica (Dogonyaro) any anti-malaria activity? *Nigerian Medical Journal, 8,* 8–10.

Ezeomah, C. (1983). *The education nomadic people. The Fulani of northern Nigeria.* Hull, Britain: Oriel Press.

Fafunwa, A. B. (1967). *New perspectives in African education.* Lagos, Nigeria: Macmillan.

———. (1974). *History of education in Nigeria.* Ibadan: NPS Educational Publishers.

Frank, S. (1975). Religion as play: Bori, a friendly "Witchdoctor." *Journal of Religion in Africa, 7,* 201–211.

Gutek, L. G. (2005). *Historical and philosophical foundations of education. A biographical introduction* (4th ed). Upper Saddle River, NJ: Merrill Prentice Hall.

Ibrahim, M. B. (1966). The Fulani: A nomadic tribe in Northern Nigeria. *African Affairs, 65,* 171–177.

International Work Group for Indigenous Affairs [IWGIA] (1998/99). Indigenous women's issues. *International Work Group for Indigenous Affairs,* 350–365.

Junaid, M. (1987). Education and cultural integrity. An ethnographic study of formal education and pastoralist families in Sokoto State, Nigeria. Unpublished doctoral dissertation, University of York, Heslington, Great Britain.

Junaidu, A. (1957). *Tarihin Fulani. Wazirin Sokoto.* Zaria, Nigeria: Gaskiya Corporations.

Kisekka, M. N. (1992). *Women's health issues in Nigeria.* Zaria: Tamaza Publishers.

Kipuri, N. O. (1991). Maasai pastoralism and the decline status of women. Unpublished memorandum. NOPA project: UNESCO/Nairobi.

Lewis, I. M., & al-Safi Hurreiz (1991). *Women's medicine, the Zar-Bori cult in Africa and beyond.* Edinburgh: Edinburgh University Press.

Mbahu, A. (1999). The traditional arts and crafts of Fulani people. *Journal of Nomadic Studies, 2,* 40–49.

Nafukho, F., N. M. Amutabi, & R. Otunga (2005). *Foundations of adult education in Africa.* Hamburg, Germany: UNESCO Institute of Education.

Ojoade, J. O. (1992). Proverbs as repositories of traditional medical practice in Nigeria. In G. Thomas-Emeagwali (ed.), *Science and technology in Africa with case studies from Nigeria, Sierra Leone, Zimbabwe and Zambia* (pp. 1–21). New York: Edwin Mellen Press.

Oliver-Bever, B. E. P. (1986). *Medicinal plants in tropical West Africa.* Cambridge: Cambridge University Press.

Palmer, H. R. (1914). "Bori": Among the Hausas. *Man, 14,* 113–117.

Ryan, K., & J. M. Cooper (2007). *Those who can teach.* (11th ed). Boston: Houghton Mifflin.

Shiva, V. (1988). *Staying alive: Women, ecology and survival in India.* New Delhi: Zed Press.

Sofowora, E. A. (1993). *Medicinal plants and traditional medicine in Africa* (2nd ed). Ibadan, Nigeria: Spectrum Books.

Stenning, D. J. (1957). Transhumance migratory drift, migration: Patterns of pastoral Fulani nomadism. *Journal of the Royal Anthropological Institute of Great Britain and Ireland, 87,* 57–73.

———. (1994). *Savannah nomads: A study of the Wodaabe pastoral Fulani of western Bornu province, northern region Nigeria.* London: Oxford University Press.

Togbega, D. D. (2006). *Traditional healers approach to malaria treatment.* Paper presented at the Africa Herbal Anti-malaria Meeting of ICAF, 20–22 March, 2006, Nairobi, Kenya.

Umar, H. D. D. (1999). Factors contributing to the survival of the Bori cult in northern Nigeria. *Numen (Lienden), 46*(4), 412–447.

Usman, M. L. (2005). Analyzing the prospects, quandaries and challenges of nomadic educational policy on girls in sub-Saharan Africa. In A. Cleghorn & A. Abdi (eds.), *Issues in African education: Sociological Perspectives* (pp. 193–218). London, UK: Palgrave Macmillan.

Veit, E., & M. Habou (1989). *Girkaa: Une ceremonie d'initiation au culte de possession Boorii des Hausa de la region de Maradi (Niger) .* Berlin: Dietrich Reimer Verlag.

VerEecke, C. (1991). Pulaaku: An empowering symbol among the pastoral Fulbe people of Nigeria. In T. Gidado (ed.), *Education and pastoralism in Nigeria* (pp. 20–25). Zaria: Ahmadu Bello University Press.

Worley, B. A. (1991). Broad swords, war drums, women's health: The social construction of female autonomy and social prestige among the pastoral Kel Faey Twareg. Unpublished doctoral dissertation, Columbia University, New York, USA.

Yusuf, K. Y. (1998). Rape related English and Yoruba proverbs. *Women and Language, 21.* Retrieved from, http://www.questia.com/googleScholar.qst?docId=5001409406

PART V

HEALTH KNOWLEDGE AND LEARNING

CHAPTER 15

TRADITIONAL HEALING PRACTICES: CONVERSATIONS WITH HERBALISTS IN KENYA

Njoki N. Wane

INTRODUCTION

To be attracted to an ancient way of life is to initiate one's personal spiritual emancipation . . . the road to correcting ills goes through a challenging path of ritual . . . To ritualize life, we need to learn how to invoke the spirits or things spiritual into our ceremonies . . . we seek strength from spirits or Spirit by recognizing and embracing our weakness.

Some, 1994, pp. 67 & 97

Our African forebears introduced us to spiritual practices and the healing legacy left to them by their Ancestors. Ancient African healing practices were founded upon holistic approaches and grounded on spiritual guidance embodied in the Creator, the giver of life, harmony, balance, cosmic order, peace, and healing. However, these practices were and are not homogenous as African societies are highly heterogeneous in nature.

According to Constantine et al. (2004) and Helms and Cook (1999), indigenous healing practices can be defined as those helping beliefs and strategies that originate within a culture or society and are designed for treating members of a given cultural group. They add that every society has designated individuals or groups of people known as healers. According to Tuck (2004), healing is a transformative process that occurs during illness, in addition to the efforts made by the healer to treat or eradicate the disease.

The purpose of this paper is to examine the healing aspects of herbalists living in seven provinces in Kenya and practicing indigenous methods to assess what type of contribution they have made to contemporary healing practices. It is based on a research project carried out in 2006 where fifty-six herbalists and thirty laypeople were interviewed. The chapter highlights only the conversations with the herbalists who have been practicing for more than ten years. The herbalists (thirty-two males and twenty-four females) ranged in age from thirty to seventy-five years old and were from such communities as Luo, Abaluya, Abagusii, Akamba, Agikuyu, Aembu, Ameru, Ambere, and Mijikenda. Most of the herbalists talked about common diseases in their communities, the herbs that they used to treat them, and how these herbs were prepared and administered. In addition, they talked about the complementarity, or lack of it, between their practices and contemporary medicine.

All the herbalists interviewed acknowledged their teachers (elders) for passing to them the indigenous knowledge, while others gave thanks to their Creator for giving them such a great gift. All healers indicated that their clientele varied as well as the rates they charged. Most patients who could not afford to pay their bills were given a reasonable schedule within which they could settle their bill on a monthly basis. In other words, an aspect of compassion was quite evident in the interviews. There was also constant acknowledgment of the various spiritual practices they employed as they sought to find the right treatment for their clients.

LITERATURE REVIEW ON KENYAN HEALING PRACTICES

In Kenya . . . little has been written about indigenous medicine and the health practices of alternative healers, but this did not stop the government from incorporating indigenous medical systems into Kenya's health system. Because of budgetary implications, the government has been unsuccessful in its mission of availing medical services to the majority of the people, and this has provided an opening for the flourishing of the alternative healthcare system.

Amutabi, 2008, p. 153

Traditional medicine remains an important primary line of health service in Kenya. It is, however, important to note that many educated Kenyans feel the use of traditional herbs is primitive and has no place in the contemporary Kenyan lifestyle. Kenya as a society has startling contrasts and striking diversity not only in terms of landscape and population but also indigenous healing practices. There are fifty-two ethnic groups in Kenya that fall mainly into three linguistic categories: Bantu, Cushitic, and Nilotic. However, despite the diversity found in this nation, there are certain common indigenous healing practices that have survived colonialism and neocolonialism and can be evoked to treat some common diseases in Kenya.

Prince and Geissler (2001) writing on healing practices among the Luos in Kenya, show the emphasis placed on relationships especially between the healers and their grandchildren. These authors point out that the quality of relationships determines what is to be learned and the nature of the pace at which the learning takes place. Indigenous practices are not preserved in books for future use, but passed on to a member of the family through observation, engaging participation, and active listening.

A study by Njoroge and Bussmann (2006) among the Kikuyu show how herbs are administered to cure malaria. Although some of the traditional herbs were tested and proven to be effective, the plants tested in the study were far too few due to the lack of comprehensive ethnobotanical data. However, the authors found that *Caesalpinia volkensii, Strychnos henningsii, Ajuga remota, Waarbugia ugandensis,* and *Olea europaea* were identified as antimalarial herbal remedies. The majority of the medicines were obtained from the roots, whereas the antimalarial species were mainly obtained from trees and shrubs.

A further study by the same authors (Njoroge & Bussmann, 2007) explored how different saps from herbs have been used to treat skin diseases associated with opportunistic infection and HIV/AIDS among the Kikuyu in Central Kenya. The researchers reported that sap or latex was applied directly to the affected areas and only in a very few instances were the plant parts boiled and the extract used for washing affected areas, usually serving as antiseptics.

However, not all Kenyans embrace such knowledge. For instance, the chair of the Kenya Medical Association, Dr. Nyikal, feels that it is immature to entrust the health of a human being to people who have no formal education and no medical training. He feels a move in that direction would take Kenya back to the ancient days. Dr Nyikal's remarks were invoked by the statement from the Minister of Health who felt the need to embrace what the herbalists could offer in terms of curing some of Kenyan chronic illnesses. The Minister stated,

> In Kenya, there has been common thinking among elites in Nairobi and elsewhere that indigenous medicine is bad and backward. This negative thinking has its origins in colonialism and persists through the minds of some. Unfortunately, the views that dominate public discussion on alternative medicine in Kenya are those of the educated elite . . . Some elites discuss among themselves in Nairobi and elsewhere and assume that everyone thinks the same way.
>
> Amutabi, 2008, p. 154

Amutabi (2008) then goes on to state: "To be sure, traditional medicine has successfully managed some ailments that had defied biomedicine in Kenya, such as asthma, diabetes, epilepsy, rheumatism, and hypertension" (p. 150). In addition, "there is no doubt that herbal medicine is working, and its popularity is far from waning despite many problems and inadequacies, such as declining availability of some herbs due to declining forest cover" (p. 158).

The sections below highlight the outcome of my research. General findings of the studies are provided, followed by the specifics of treating common diseases found among different ethnic groups in Kenya.

GENERAL FINDINGS

The findings of my research presented reveal that indigenous healing practices have survived in spite of popularization of Western medical practices.
In addition, these practices not only involve work that corrects the internal
imbalances through which disease can manifest within an individual, but also
emphasize the reestablishment of individual harmony with the environment
and the relationship with the natural cycles to which all life is subject. The
findings from this research have shown that traditional healing practices take
into consideration the physical, emotional, mental, and spiritual realities of
a person seeking help. The services provided by herbal doctors are unique
and holistic in nature as they take the patient's total self into consideration
during the treatment sessions. As more and more people begin to embrace
alternative modes of healing, this aids mainstream thinking by incorporating
alternate knowledge into nonholistic approaches, while concurrently aiding
in the rebirth of self-dignity and self-worth among the practitioners of the
more traditional approaches.

Attempts were made by the participants to define some terms relevant to
the indigenous healing practices. These definitions gave an understanding
of the respondents' initial perspective of indigenous healing practices. About
50 percent of the participants defined indigenous healing practices as treatment
and management of illness or disease using herbs or traditional methods
passed on from one generation to another without the use of modern means
of treatment, prevention, or management, and 70 percent of the respondents
reported having acquired this knowledge from older members of their societies
while some (10%) reported having acquired the knowledge from family members or friends and 4 percent indicated that they had gotten the gift from God
or the Creator. Furthermore, 20 percent of the participants reported getting the
knowledge through interaction with known herbalists and from contemporary
reading on herbal medicine. It was interesting to note that most (84%) of the
participants interviewed reported that knowledge regarding indigenous healing practices is not common to all members of the society. They cited lack
of interest by most people, due to a shift in the reliance on modern medicine.
In addition, they felt that in order to be a herbalist, one had to be trained by
a reputable herbal doctor. Of all the people who were interviewed, 99 percent
agree with the existence of traditional healers in their communities and all the
interviewees believe that the traditional healers have the knowledge and skills to
cure most diseases. The majority of the participants (98%) reported that there
were between eight and ten doctors in their communities. Only 45 percent
attempted to define the use of local methods in treatment of diseases based on
local values and beliefs with the use of local plants.

Some of the herbalists were more established than others. Some sold their
herbs at the local markets while other treated their clients in their homes.
With the local market herbalists, if any of their clients wanted a follow-up,

the clients would either return to the healers at the market or go to the healer's home. These herbalists sold their herbs either in liquid form or as roots, barks, or leaves and the patients were given instructions on how they should prepare the herbs. Most times, they were instructed to boil them and drink the concoction until they felt better. If the patients bought the already prepared herbs (it was usually in one-liter bottles), they would be instructed to drink all the medicine and to come back for more if they did not get better.

SPECIFIC FINDINGS

Although indigenous healers vary by culture and type of healing, my research revealed that some healers undergo rigorous training and education in order to attain their positions. The training is often transmitted, learned, and remembered in an oral, nonliterate tradition. For instance, among the Luo, Prince and Geissler (2001) describe a close relationship between a Luo healer and her grandson, who learns not only the healing practices but also social skills such as respect and compassion for people. Prince and Geissler explained that

> the passing on of medicine and medical knowledge establishes a lasting bond between the giver and the recipient. The gift of medicine from grandmother to grandchild thus contributes to the closeness of their relationship, which will last beyond the grandmother's death. (p. 461)

Thus the selected grandchild becomes the designated family member to continue with the tradition as healer, and will spend much of his time with the healer, observing and learning how to harvest, prepare, and use the plants to treat various sicknesses. Prince and Geissler's (2001) finding is similar to what some herbalists in Meru, Embu, and Kisumu explained in interviews during my research. Among them Mugwimi, a Meru herbalist:

> I acquired most of what I know from observation. I was very keen to learn from my grandfather, however, I also took courses in herbal healing and that is my diploma (pointing to where he had displayed a framed certificate). My reason for following my grandfather's footstep is due to frustrations of diseases that are resisting contemporary medicine and seeing patients die in large numbers in hospitals where I work as clinical officer.
>
> Interview, 2006

While Mwene, a herbalist from Embu, stated:

> I was intrigued by the way my father used to use herbs to treat his family members and his animals. I got interested and wanted to know more and

I started to follow old men from my village to the forest in search for roots and herbs. I would ask them the names of different plants, and their purpose. I then researched in books and when I completed high school, I continued with this search . . . Since I wanted to make sure what I was doing was not going to kill people, I used to take any herbal medicine that I would prepare to KFRI (Kenya Forest Research Institute) for verifications. I would then consult more books in order to know what to mix with what and what dosage to administer.

                                                                    Interview, 2006

Dr. Kelly, another Meru herbalist observed:

I owe what I know today to my grandfather and my father. I have been observing them since I was a small boy . . . and I know from their work and the work that I do now, there are many and new untreatable diseases in Kenya, and modern medicine has failed to find a cure . . . Many people come to our clinic and we feel helpless sometimes, however, this helplessness has made us to carry out more research and to come up with cures for some of the new diseases.

                                                                    Interview, 2006

Jelo, a Luo herbalist stated:

I acquired this knowledge from my father when I was young. I feel it is my responsibility to pass this knowledge to my children. I usually take my children to the forest when I go in search of medicinal plants.

                                                                    Interview, 2006

Prince and Geissler (2001) however notes that most of the rural population has some form of indigenous healing:

[A]ll practitioners of Luo medicine, children and adolescents, mothers, grandmothers, and well-known herbalist-healers, share a common body of skills concerning illness and herbal treatments. This is learned as part of growing up in rural homes, where mothers and grandmothers use herbal medicines in dealing with children's illness. It also is learned through relationship with grandmothers. What a "healer" knows and does is therefore rooted in communal ideas and practice." (p. 454)

This observation was also noted among the herbalists I interviewed. For instance, a Nandi agricultural officer who had grown up in the city learned about indigenous medicine from lay people. He stated, "I learned all what I know today from different farmers while I was working as an agricultural officer among the Nandi people. Most of these farmers know how to treat common colds, stomachaches, cuts, snake bites." Kiene, a Kikuyu herbalist, learned the practice from his father who was not a herbalist:

> My father was not a herbalist, but he had a wealth of information about plants, roots, leaves, or even barks of trees. As we went grazing our cattle, my father would uproot some plants and tell me to chew, while others were taken home and given to my mother to add to soups and everyone in the house was made to drink the soup. This made me interested in finding out more about the medicinal value of plants.
>
> Interview, 2006

Jadong, a Luo herbalist stated that she learned about herbal medicine from talking to men and women in her village about curing diseases.

Box 15.1 lists the diseases that the herbalist could treat.

Box 15.1   Diseases treated by herbalists

typhoid; malaria; arthritis; amoeba; pneumonia; STDS; TB; dysentery; high blood pressure; stroke; rheumatism; brucellosis; HIV/AIDS; diabetes; bronchitis; dog bites; snake bites; asthma; backaches; ulcers; allergy; prostrate cancer; kidney problems; loss of memory; fibrocystic breast condition; heartburn; infertility; measles; cancers; pneumonia, eczema, gout, eye diseases, loss of appetite; tonsillitis; epilepsy; teeth problems; toxins; kidney problems

The following are just a few of the names of plants or herbs that the respondents gave:

Box 15.2   Names of plants and Herbs

Blackjack; Rayudhi and Sede; Osol Olaw Otange, Osiro, Ochieng kanyadho (no translation); ochuogo; onyalo biro; okita; akeyo dek; dwele; ogaka olandra; rayudhi (Luo names). haloulu, khasanga riuba, nasiumnya, mihweso; muhululo; sigomori; induli and osangula (Luhya names); mwalula (herb); kinondo (herb); *sodoapple* (fruit); muteta (root); mukaakaa (root); muthulu (root); mwola ndathe (root); Kitungu (root); kikuni (root); kithiiya (root); kirurite—*Tithonia diversifolia* or *Amarathus Grassilas*; kithare—Dracaena Steudreri; kitherema—lannea; mburu—*Vitex Payos*; mubebu—trema orientalis; muchugu—*Cajamus Cajan*; muchunguchungu—*Crotalaria Axillaris*; mugucua—*Zanthoxylum Chalybeum;* mugumo—*Ficus Thonningii*; mukawa—*Carissa Spinarum*; mekeu—*Dombeya burgessiae*; mukinduri—*Croton Megalocarpus*; mukunguu—*Earythrina Melanacantha*; mukuu—*Ficus Sycomorus*; munyanwe

The following information provides an overview of how different ethnic groups treat and prevent malaria, which is a very common disease among most communities in Kenya and has claimed many lives. Malaria is caused by a plasmodium or a protozoan transmitted by a female anopheles mosquito usually found in stagnant water, open pits, open containers, or overgrown shrubs. Marium, an Akamba herbalist, described how to diagnose a patient suffering from malaria:

> The sick person will complain of joint pains, headaches and sometimes dizziness, vomiting, diarrhea, shivering and sometimes fever . . . I use *neem* products. The treatment involves boiling leaves or a piece of bark for five to ten minutes, and then the mixture is sieved and left to cool (the solution may also be taken warm). Because it is bitter, honey is added. The adults are given one glass of the boiled solution three times a day and children should be given a quarter glass three times a day . . . always after meals, until the person feels better. In addition, once the leaves or the bark of *neem* are boiled, the sick person may be covered with a blanket for ten to fifteen minutes under the container with the boiled leaves or bark so that they can inhale the steam to facilitate sweating. In addition, the patient is bathed in some of the water. The process of inhaling and bathing may be repeated twice a day until the patient feels better. This is because it is believed that when the person inhales the steam, the malaria is healed from inside as well as outside the body.
>
> Interview, 2006

Jeff, a Kisii herbalist, explained how some Abagisii people who were suffering from malaria were easily cured by following any of the treatments explained below:

> A. A goat is slaughtered and contents of the small intestine are boiled together with the meat and the patient is then given the soup to drink . . . twice a day . . . We believe that a goat eats all sorts of herbs which are believed to be medicinal.
> B. Or a patient may be given a drink prepared from *Omobeno* roots—the drink makes the patient to have a diarrhea. The diarrhea is meant to remove the protozoa from the stomach . . .
> C. Or the patient is given the content obtained by boiling *Omoutakiebo* leaves. The drink makes the patient to vomit. In inducing vomiting, it is believed, the malaria parasites are vomited out. This is repeated twice a day.
>
> Interview, 2006

Among the Maasai, malaria was treated by making a patient inhale steam from boiled water with *msanduku* (a plant). As with the Akamba treatment, the Maasai may also boil leaves or a piece of bark, then sieve the mixture and leave it to cool, and then given to the patient. Because it is bitter, honey may be added. This solution is given three times a day, until the person feels better. Malaria may also be treated using grapefruit, which is believed to contain quinine. According to Leleso,

> the grapefruit is boiled, and then mashed, strained or sieved and the liquid is then given to the patient three times a day for five days. If the patient is experiencing nausea or vomiting, he or she may be given a mixture of salt and lemon to lick.
>
> Interview, 2006

Another Maasai herbalist, Tunanai, explained that

> malaria among the Maasai can be treated using *entipilikua* plant whose properties are boiled for about thirty minutes to an hour, until the water turns reddish-brown in color. Once the solution is cool, the patient is given half a glass twice a day . . . *Olg'osua* thorns can be used to treat malaria. The thorns are boiled then the patient is made to breathe the vapor, while covered with a blanket, and this makes him/her sweat. The steam relieves the running nose. This is done once a day.
>
> Interview, 2006

Aloe Vera is another plant suggested for treating malaria among most ethnic groups. According to most responses, the preparation required is similar to that of the *neem* tree; the only difference is that aloe vera is added to water that has boiled and been left to cool for two hours. The water changes to a green color. Adults should be given a glass of the solution three times a day and children given a quarter glass three times a day. Both *neem* and aloe vera are very bitter solutions and the herbalists suggested that honey should be added to medicine.

Among the Embu people, malaria could be treated using the following herbs: mukambura; mufa, njeru warurii, mucuthi, thuthiga, mwiria (*rosaceae or prunus africana*); muvuti, mogoran (*rhamus prinoldes*); muguucwa (*zenthoxylus asiatica*); mucharage (*olea capensis*); mujuthi (*caeslpiniaceae*); munganga, kirurite (*lamiaceae*); muchani; murao; mukinduri, muthiamura; (*scena; periodic*); muthiga; and (*cinchona*).

Most Akamba, Aembu, Ameru, Agikuyu, Abagugii, and Abaluya families have been advised to plant the midnight queen bush as a preventative measure for malaria. The plant has a very strong and sweet scent, which repels mosquitoes; however, it is not used as a curative herb. Another preventative measure that communities are encouraged to follow is to clear all overgrown shrubs within the vicinity, burn all used containers, fill in all open pits, and pour kerosene on all stagnant waters.

Toothache was another common disease that many herbalists talked about. According to Mureri, an Embu herbalist,

> you can treat a toothache using a dry avocado seed . . . crush the seed to produce fine flour. Take the fine powder and put it in the mouth and wait. The

pain will go away after some time . . . *Muti wa miago* (fuerstia Africana) could
be used to treat toothache.

Interview, 2006

Other methods of treating toothache included use of cinnamon, *schimus
molle* (which is also used in treating a variety of wounds and infections),
devil's horsewhip, or asparanthus aspera plants. The herbalists advised that
with the cinnamon, one could put some in their tooth, while with schimus,
the milky sap from its leaves is used. Some herbalists advised their clients to
brush their teeth twice a day using particular herbs such as devil's horsewhip
or asparanthus aspera.

A Luo herbalist explained that toothache could be treated by chewing
the leaves of Osol olaw (Luo) then spitting out the juice after a while, while
among the Akambas, Itongu (*Sodom apple*) was a common cure for tooth-
aches when its juice was mixed with cold water, then gargled and spat out.

Stomachache or diarrhea were ailments that many of the herbalists spoke
about. Mabilia, a Luhya herbalist, explained:

> There are different methods of treating stomachache or diarrhea. For instance,
> aloe vera cut into small pieces then soaked in hot water for two hours is a good
> cure for stomachache or diarrhea . . . The adults are encouraged to drink one
> glass of the solution three times a day and for children a quarter of a glass.

Interview, 2006

Vero, another Luhya herbalist, suggested use of *haloulu* or *khasanga riuba,*
which must be uprooted properly then boiled. The content is then stored
in a bottle and the patient is given small portions of the medicine. Another
Luhya herbalist explained that diarrhea could be cured by using omusengese,
which could be boiled well and half a glass given to the patient twice a day,
morning and evening. Among the Luos, diarrhea was treated using ng'owo
poko. In addition, akeyo (Luo) leaves could be used to treat stomachache,
while among the Akambas, stomachache could be treated using kinondo
(herb) mixed with kivu (just the roots), kitungu (both its roots and herbs were
used), and kiluma.

Ulcers, another common stomach problem, according to Regitari, a herbalist from Embu can be treated following the following method:

> Take a cup of milk three times a day or alternatively you can cut the euphobia
> tree indigenously called muriaria in the Embu language. Split the plant and
> peel it, chew the peeled plant and swallow. For this process to be effective, one
> has to follow it for one year. This leads to an everlasting cure of ulcers.

Interview, 2006

According to Mujai (a Luo herbalist), solanum nigrum which is known
as *osuga* in Luo is very effective treatment of ulcers, while the Merus use

kariaria (an herb in Meru language). Unfortunately, these herbalists did not want to explain how they used these herbs to treat ulcers.

Typhoid is a very common disease especially during the rainy season. According to herbalists that were interviewed, typhoid is caused by symonella typhi usually found in dirty water and is transmitted through drinking untreated water and eating food that is not well cooked. The disease usually comes with similar symptoms to those of malaria such as: joint pains, headaches, vomiting, diarrhea, shivering, and sometimes fever and confusion. Mariam, a Luo herbalist, explained that typhoid can be treated using aloe vera, cutting the plant into small pieces and mixing these with boiled water until the water turns green. After two hours, the herbalist adds some honey and the solution may be taken. Typhoid, according to Mariam, could also be treated using the *Olandra* (Luo) root, which is dried, pounded, and one teaspoon mixed with hot water. The patient is advised to drink half a glass of the solution twice a day.

SOME DIFFERENCES BETWEEN INDIGENOUS HEALING AND CONTEMPORARY PRACTICES

Some interview respondents said that in the past, people used to fear herbalists as they were referred to as witchdoctors. However, there is now a greater acceptance, as there are many well-known herbalists who have clinics all over the country. One respondent said

> Herbalists have opened up these clinics in good time when lots of diseases cannot be treated with modern medicine . . . Some herbalists even counsel and teach the patients on how to keep health. . . . And when the patients practice what they are taught, they do not fall sick

Interview, 2006

Dr. Kelly (herbalist) said that those who have knowledge of herbal medicine could share it with contemporary practitioners. He felt there was a need to enrich each other's professions, explaining that most of the herbalists do not have equipment for packaging their powdered medicine or even X-ray machines, and if the two professions worked together, they could provide support to each other. Dr. Kelly gave examples of some hospitals that referred patients to him or even called him to visit patients in hospital wards. He recommended that herbal medicine be regularized and standardized so that the patients are given similar treatments across the country. Noni, a herbalist, also believes that both practices can complement each other. For example, she wished at times that she could know exactly what a patient was suffering from and imagined that if she did, it would be easy to administer the right herbs. She emphasized the need for laboratory tests for her patients, explaining that she always sent her patients to a medical doctor for tests or stitching of open wounds if that was required. Mweni, a herbalist added: "I long for that day when we can sit around the same table with medical doctors and discuss some of the common diseases that are killing our people."

Most of the herbalists do not have proper equipment to prepare the medicine and as a result it can be difficult to carry it around. Contemporary medicine has different ways of diagnosing the sickness; with herbal healers, it could be guess work sometimes and as a result, there is a need to work together to complement each other. Herbal medicine is in its natural state, it has no chemicals, and no side effects, and it provides nutrients that kill the bacteria in the body. Some of the major differences are the forms of sterilization: herbalists use lemon, salt, garlic, and charcoal, while contemporary practitioners use sterilization equipment or different forms of liquids. With contemporary medicine, there are laboratories where samples of blood or urine are sent for diagnosis, while traditional practices rely on the results brought from the hospital, hence a lack of facilities interferes with their work.

Mure, a herbalist explained:

> Most of us have helped patients who have suffered for many years even after going to hospitals, so if the government recognized the work we are doing and provided laboratory services for us, they would enable us to advance our work . . . Most patients come to see me because the hospitals have failed to cure their illness. I, however, send patients to hospitals to carry out laboratory tests, and once I confirm their sickness, I start treating them. I cannot treat patients without their medical history from their doctors or from hospitals where they have been attended to.

> Interview, 2006

Rutei, a herbalist, said in an interview: "Both types of practices can be used, for example, use of plasters for broken bones is good . . . we do not have the equivalent of that in herbal medicine."

DISCUSSION OF THE FINDINGS

African indigenous healing practices are curative and used by traditional African societies and still being practiced today. Ancient African indigenous practitioners taught the patients about herbs and healing (Afrika, 2004). Using this mode of knowledge dissemination, almost all families acquired basic knowledge of health and healing. Many lay people who were interviewed in my study and who used herbal medicine mentioned this. Thus, every family should learn the basic methods for treating common diseases. Unfortunately, most of the herbalists interviewed guarded their secrets and did not share specific methods of preparation. The information they provided was very generic and it is not advisable for people to try the methods reported above without proper training.

In African nations, the fear of reprisals by governments has caused the decline in traditional healing practices and has forced most people to do their healing work very quietly, within a circle of trusted supporters. However, with the resurgence of ethnic practices around the world, traditional healing practices are gradually gaining strength and are making a slow but sure return. There is a need to reclaim this knowledge, to revive the connection to one's roots, and to acknowledge that traditional healing is timeless; it is both an ancient medicine and the medicine of the future.

Traditional healing sees the universe as operating according to the laws that regulate wholeness and balance. Among many African peoples, there is a strong belief that the purpose of life is to strive for balance and this cannot be achieved without knowledge of natural laws and their corresponding elements of earth, fire, water, and air. It is therefore important to begin healing by reconnecting with the elements of nature, because nature is "our first home, the foundation of our community, the dwelling place of the spirits who watch over us and long to be reconnected with us" (Some, 1999, p. 57).

Healing is also about being whole (Graveline, 1998; Hibbard, 2001; Shilling, 2002). It does not enhance one's domination over others, but promotes harmony and balance not only among human beings but also among nature, animals, and the supernatural. Peter Reason (as cited in Shilling, 2002) states: "To heal means to make whole: we can only understand our world as a whole if we are part of it: as soon as we attempt to stand outside, we divide and separate" (p. 156).

Relationality as emphasized by Luo herbalists is also important because it maintains balance and harmony through healing. Relations are formed with the natural environment, family, community, and others. This is an important aspect of healing because relationships not only involve correcting internal imbalances through which disease can manifest in an individual, but emphasizes reestablishing an individual's harmony with their environment and their relationship with self, family members, and the community as a whole.

CONCLUSION

Traditional healing practices, when compared to Western health systems, are viewed as distinct systems with a different model of diseases operating within a different worldview. In the Western healing model, the healing process is initiated when the doctor gives the patient some medication to deal with the sickness. In traditional healing practices, the rituals are more or less the same. The patient is prescribed some herbs to take. In addition to that, the healer provides prescriptions for healing the root cause of the sickness. In the analysis of illness in contemporary times, there is an assumption that the patient's sickness is a separate entity from the domains of social, economic, or cultural structures. Among the traditional healing rituals, there are implicit questions such as: What has disrupted the circles of balance in one's life? What will it take to bring harmony and flow of energy back to its acceptable body or environmental normative? What type of preparation is necessary for complete healing? Although these questions were not on the interview guide, they were addressed very eloquently by most herbalists who indicated that, in addition to healing the particular sickness, it was important to establish the root cause of the disease. For instance, if the patient was suffering from malaria, the herbalist would advise the client to clear the bushes around the home and if possible to buy a mosquito net. If the patient was suffering from *bilharzia,* or any water borne diseases, they would be told to treat the water or to find ways of getting clean water for home consumption or they would be advised to wash their hands, vegetables, and utensils thoroughly. If related to any form of cancer, the herbalist would explain to the patient the cause of the disease. As a result, the patient would be fully aware of what was being treated and how they could avoid recurrence of that sickness in future.

Traditional and contemporary practices should be married together as each has its strong points and complement the other. Dr. Kelly said: "Contemporary healing is more advanced; however, this healing is some-times not effective at all. I think indigenous practices are better in some aspects and contemporary practices have their strong points as well." Indigenous practices should be encouraged and recommended as comple-mentary to contemporary practices, especially since people of all back-grounds use both practices.

To begin to make sense of healing practices in Africa, one needs to understand the worldview of the African people and the complexities of each culture. From the work I have done, there is clear evidence that there is a need for more research on ethnomedicine. The term traditional healer refers to a person involved in one of a broad range of practices. Knowledge is based on practices, explicable or not, used in diagnosing, preventing, or eliminating physical, mental, or social disequilibrium, and relies on experi-ence and observation handed down through generations.

Acknowledgment Note: I would like to thank all the research assistants from Catholic University in Kenya: Ursulla, Catherine, Joyce, Paul, Itaclaria, Florence, Maina, Nyaga, and Anne who participated in this work. I will always be grateful for the funds from Catholic University of Eastern African, Kenya to carry out this project. I would also like to thank Professor Ogula for being my co-researcher in this project. Last but not least, I would like to thank Yumiko for helping gather the literature on Indigenous healing practices.

REFERENCES

Afrika, L. O. (2004). *African holistic health.* Brooklyn, New York: A & B Publishers Group.

Amutabi, M. N. (2008). Recuperating traditional pharmachology and healing among the Abaluya of Western Kenya. In F. Toyin & H. M. Matthew (eds.), *Health Knowledge and Belief System in Africa* (pp. 149–170). Durham, Carolina: Academic Press.

Constantine, M. G., L. J. Myers, M. Kindaichi, & J. L. Moore III (2004). Exploring indigenous mental health practices: The roles of healers and helpers in promoting well-being in people of color. *Counseling and Values, 48*(2), 110–125.

Graveline, F. J. (1998). *Circle works: Transforming eurocentric Consciousness.* Halifax: Fernwood Publishers.

Helms, J. E., & D. A. Cook (1999). *Using race and culture in counseling and psychotherapy: Theory and process.* Boston: Allyn & Bacon.

Hibbard, P. N. (2001). *Remembering our ancestors: Recovery of indigenous mind as a healing process for the decolonization of a Western mind.* San Francisco: California Institute of Integral Studies.

Njoroge, G. N., & R. W. Bussmann (2006). Diversity and utilization of anti-malarial ethno-phytotherapeutic remedies among the Kikuyus (Central Kenya). *Journal of Ethnobiology and Ethnomedicine, 2*(8), 1–7.

———. (2007). Ethnotherapeutic management of skin diseases among the Kikuyus of Central Kenya. *Journal of Ethnopharmacology, 111*(24), 303–307.

Prince, R., & P. W. Geissler (2001). Becoming "one who treats": A case study of a Luo healer and her grandson in Western Kenya. *Anthropology and Education Quarterly, 32*(4), 447–471.

Shilling, R. (2002). Journey of our spirits: Challenges for adult indigenous learners. In E. V. O'Sullivan, A. Morrell, & M. A. O'Connor (eds.), *Expanding the boundaries of transformative learning: Essays on theory and praxis* (pp. 151–158). New York: Palgrave.

Some, M. (1994). *Of water and spirit: Ritual, magic, and initiation in the life of an African shaman.* New York: Teacher/Putnam Books.

———. (1999). *The healing wisdom of Africa: Finding life purpose through nature, ritual, and community.* New York: Teacher/Putnam Books.

Watson, J. (2003). Preparing spirituality for citizenship. *International Journal of Children's Spirituality, 8*(1), 9–24.

"TO DIE IS HONEY, AND TO LIVE IS SALT": INDIGENOUS EPISTEMOLOGIES OF WELLNESS IN NORTHERN GHANA AND THE THREAT OF INSTITUTIONALIZED CONTAINMENT

COLEMAN AGYEYOMAH, JONATHAN LANGDON, AND REBECCA BUTLER

INTRODUCTION

In a recent article, Dr. Peter Arhin (2008), Director of the Traditional and Alternative Medicine Directorate of Ghana's Ministry of Health (MoH), wrote, "[T]raditional medicine and complementary and alternative [medicine] is emerging from a long period of systemic global marginalization" (p. 4). Arhin is not alone in placing a renewed emphasis on Traditional Medicine in the Ghanaian and African contexts. In fact, the World Health Organization (WHO) has noted, "TM [is] one of the surest means to achieve total health coverage of the world's population" (Sarpong, 2008, p. 6). This is especially true in Ghana where 70 percent of the population use traditional medicine as its first choice for well-being (Arhin, 2008; Sarpong, 2008).

Yet, tempering these accounts of the renewed recognition of TM (traditional medicine) in Ghana and Africa is a cautionary note about the "safety concerns" and much needed "regulations" associated with this alternative "system" of medicine (New Legon Observer, 2008, pp. 1–2). In order to address these safety and regulatory concerns, the solution to many is "integrat[ing] [traditional and] herbal medicine into the mainstream of national healthcare systems" (New Legon Observer, p. 1). One key piece in addressing these concerns, according to Arhin, Sarpong, and other proponents of this point of view, is an institutionalized formal education system for traditional medical practitioners (TMPs).

At the same time, with due respect to these integrationist opinions, one has to wonder, with so many Ghanaians using the "traditional medical system," if it isn't the allopathic system that needs adjusting. This contrasting opinion was evident in a recent community based assessment of health, where an elder of one community stated "to die is honey, and to live is salt," meaning he would rather die without using the allopathic system than use it and live (Gariba & Langdon, 2005). For this elder, the problem is not the "traditional medicine system," but is rather the presence of the allopathic system—a complete inversion of the concerns expressed above.

In this chapter, we present an interpretation that takes the opinion of this elder seriously, and therefore runs contrary to the arguments of Dr. Arhin and Dr. Sarpong, as well as the WHO. In fact, our argument is so contrary that we do not even use the language that these others have used to talk about the alternative approaches to wellness that are often mislabeled and lumped together as traditional medicine. Key to this counter argument is a rearticulation of wellness not as a pathological practice, but as philosophy of life (Agyeyomah & Langdon, 2009). To discuss traditional medicine as merely a practice is to ignore these major epistemic issues and only underscores the reason why the systematic approach of orthodox Western medicine is incapable of evaluating the strengths of this alternative path to wellness. We have made these arguments elsewhere (Agyeyomah & Langdon, 2009) in a more general sense. In this chapter, we focus specifically on the dangers that current plans to institutionalize the training of what the WHO calls TMPs in Westernized educational contexts pose not only to the wellness approach most Ghanaians prefer, but also to the underlying philosophy behind this approach. In order to explore the full ramifications of these two dangers, as well as articulate a holistic alternative formation process, we conclude the chapter with an extended discussion of the philosophy of wellness and learning of Chief Isshaku Gumrana Mohammadu, a bonesetter in Northern Ghana. Our conversations with Chief Isshaku have been ongoing for the past four years and the complex vision of wellness that has emerged from these conversations has been deepened through a process known as narrative restorying (Randall, 1996). This chapter includes restoried reflections garnered through discussions held in March 2009.

RESPECT THROUGH INTEGRATION? THE COLONIZATION OF ALTERNATIVE PATHS TO WELLNESS

We have highlighted the problem of language above intentionally, for this problem is wider spread than the mere use of such misnomers as TM, and TMPs. For instance, is it not inherently problematic to call a health system *mainstream* when its impact is felt by only 30 percent of a population, when only 30 percent of a population believe its approach to wellness is the key to health? There is clearly a politics to this, grounded in a convergence of knowledge and power that enables this kind of stark absurdity to survive scrutiny. In effect, this politics of health can only make sense through the lens of what Walter Mignolo (2000) has called the coloniality of difference—where local ways of knowing and being are marginalized by the global designs of other local ways of knowing. Viewed through this lens, the mainstream health system Arhin and Sarpong speak of is not mainstream because it is used by the majority of Ghanaians, but rather because it dominates the public imaginary through its connection to the colonizing force of Western allopathic medicine, which in turn is linked to neoliberal globalization.

In a sense, then, when Arhin (2008) writes that "traditional medicine and complementary and alternative [medicine] is emerging from a long period of systemic global marginalization" (p. 4), he is referring to the marginalization of a wide array of alternative paths to wellness in the public imaginary by the constant message that Western medicine is true medicine, and to refuse it is to be backward. However, there are a number of problems with this—not least the fact that even in Ghana's current state as an African neoliberal good news story (Carothers, 2002), 70 percent of Ghanaians deliberately refuse the allopathic path and choose a path with deeper local roots. Even when we add the corollary argument that notions of public imaginary often come hand in hand with literacy (Anderson, 1991), this situation still does not fit. For instance, Ghanaians living in urban areas as well as those "domiciled abroad" use local wellness systems—this is neither a phenomenon of poverty nor of lack of Westernized education (New Legon Observer, 2008).

Following from this obvious inversion of what is the chosen mainstream path to wellness in Ghana, the supposed reemergence of "traditional medicine" from its "marginalization"—and those who celebrate it—must be questioned further. Four such authors/groups of authors will be closely read here: Arhin (2008), Sarpong (2008), Onuminya (2004), and Aïres, Joosten, Wegdam, and van der Geest (2007).

Arhin (2008) makes a strong case for the need for legislation in connection with what he calls "Traditional Medicine and Complementary and Alternative [Medicine] (TM/CAM)" (p. 4). This is because:

> Legislation. . . . remains the only sure way of legitimizing policy and officially recognizing TM/CAM. This is more likely to induce sustained commitment to

providing structures, funds and logistics for a visible and measurable implementation process. (p. 5)

Not only will this legislation help recognize and protect TM/CAM, it will also
help promote its "rational use" (Arhin, 2008, p. 6). In Ghana, the "Traditional
Medicine Practice Council Act (2000) . . . makes provision for the registration, regulation of standards of practice of TM practices and the setting-up
of a regulatory council with administrative offices throughout the country"
(Arhin, 2008, p. 5). For Arhin, with this regulatory framework in place, he
can envisage a time in the future when Ghana's TM practice can become part
of the modern medical system, as it is in such places as "China, India and
Germany" (p. 6). Key to this integrative process is "a comprehensive training
programme for selected indigenous TM practitioners . . . which is ready to
be rolled out" (p. 6)—a statement one must take seriously given the author's
high rank in the MoH. Yet, nowhere in this vision is there any respect for the
local epistemology of wellness. For instance, the web of regulatory practices
that Arhin sees being put in place with regard to this complex arena of wellness is being planned as if there have never been any indigenous processes of
regulation, the most potent of which is the long formation process of practitioners (see bonesetter discussion below). From this critical perspective, the
regulatory framework Arhin describes takes on a more sinister and colonizing
nature, as the discursive web through which the "mainstream allopathic health
care system" hopes to subjugate, contain, and control what, to the majority of
Ghanaians, is clearly a real alternative approach to wellness.

It is not by chance that Arhin (2008) mentions training as the main
mechanism through which this process of regulation and containment will
occur. In effect, this is one of the main commonalities of the vast majority of
writing on TMPs and bonesetters by allopathic researchers (see for instance
Eshete, 2005; Hoff, 1992; Onuminya, 2004; 2006; Sarpong, 2008; Shah et
al., 2003). Of these, Kwame Sarpong (2008) bears the most relevance on the
current topic, as he is director of a new university level training program for
traditional medicine practitioners in Ghana. Sarpong (2008) notes the training program is based on the recognition by the WHO that "it is difficult to
achieve any health delivery without the active participation of TM" (p. 15).
This goal led the WHO to adopt "strategies for developing and institution-
alizing traditional medicine as part of health care delivery" (Sarpong, 2008,
p. 16). Key to this strategy is the "training and registration of Traditional
Medicine Practitioners to be gatekeepers in health promotion" (p. 16). As a
result of this strategy, Ghanaian MoH officials have begun training programs
for Ghana's "approximately twenty thousand (20,000) Traditional Medicine
Practitioners (TMPs)" (Sarpong, 2008, p. 16). Yet, according to Sarpong,
they face a challenge: "a large number of TMPs are ageing, illiterate, lack
scientific training and lack competency in diagnosis" (p. 16). In order to face
this challenge "a four year degree programme in herbal medicine" is planned

"as a major step towards the development and advancement of traditional medicine" (p. 16). Yet while the course description provides for a "solid foundation in biomedical sciences" there is absolutely no recognition of the philosophy of wellness behind the practice of what Sarpong and the WHO call TMPs. Additionally, despite its lesser status as the "healthcare system" of choice for Ghanaians, there is no provision for a course that educates allopathic practitioners on the philosophy and practice of these pejoratively labeled TMPs. This curricular approach reveals the colonizing logic behind this training. Given this mindset, it should not be surprising that bonesetters and other alternative practitioners have such misgivings around these types of training programs (Ventevogel, 1996).[1]

In bonesetting research we can see a similar colonizing mindset, even in the case of those attempting to report on bonesetting practice fairly. For instance, Onuminya (2004) presents the results of a five-year "participatory documentation" of bonesetting practices in Nigeria (p. 653). He begins his report by extolling the potential for "teamwork among all categories of health workers" (p. 652). Through this teamwork, he suggests, traditional and allopathic practitioners will learn from one another and this will lead to mutual respect. Continuing on this model of teamwork, he describes how, as part of his approach to the participatory research, "mutual rapport was established with the chief TBS [traditional bonesetter]" (p. 653). However, the researcher's respect for these chief bonesetters is only surface-deep. For instance, elsewhere he has written, "In developing nations traditional bonesetters (TBS) play a significant role in primary fracture care. However, despite high patronage TBS remains [sic] an untrained quack whose practice is often associated with high morbidity" (2006, p. 320). He further suggests that belief in the bonesetter's practice by rural Nigerian communities is misguided (2004). His proposed solution to this misguided trust is similar to Sarpong's call for training. "I believe," writes Onuminya (2004), "that TBS practice could be modified and improved to make it appropriate in the management of carefully selected fracture cases" (p. 657). In order to achieve this, he argues, "the TBS should be further trained and integrated into PHC [primary health care] delivery, especially in primary fracture care services" (p. 657). This is yet another example where an author embedded in the allopathic tradition can only understand learning in one epistemic direction, despite arguing for the need for "teamwork." The further dismissal of the trust of Nigerians in the healing processes of the bonesetters is equally problematic, especially if ones purports to see bonesetters as an important part of primary care. When this clear bias is coupled with Onuminya's (2004) failure to pay any attention to approaches beyond the practical—only briefly mentioning the fact that fees for service "are paid at the discretion of the patients" (p. 653), and ignoring the philosophical reasons behind this approach—we clearly see the lacuna of allopathic practice manifesting itself. It can only be sheer egoism that has led allopathy to be incapable of conceiving of another approach as preferred and of this preference being based on experience and respect rather than ignorance. Interestingly, what struck us as we read Onuminya's (2004) piece is not his typification of Nigerian bonesetting as problematic and backward, but rather how similar the approaches he describes are to those used by our own guide into this world, Chief Isshaku. Not only is there a similar approach to fees in locations 700 kilometres apart, but the photos in the article suggest very similar splint-binding techniques. This indicates a wide range of knowledge sharing that has been ongoing for many centuries—something Chief Isshaku confirms below—and helps further the case for thinking of indigenous and traditional health knowledges as dynamic, evolving, and widespread systems of knowledge sharing, rather than as the static entities Western thought has for so long sought to colonize (McMichael, 2004).

In contrast to Onuminya (2004), Aïres et al. (2007) present a case that takes Ghanaian choices about healthcare seriously. Set in Techiman, in Ghana's Brong Ahafo region—roughly 300 kilometers from Chief Isshaku's village—this study returns to an area where previous studies have been conducted on bonesetting in particular, and TM in general (c.f. Ventevogel, 1996; Warren, 1974; Warren, Bova, Tregoning, & Kliewer, 1982). This study focused specifically on the question of bonesetter effectiveness from the vantage point of patients rather than from that of Western trained clinicians (Aïres et al., 2007). The study revealed much the same attitude that our work with Chief Isshaku has shown (Agyeyomah & Langdon, 2009): bonesetters are held in high esteem by local populations, not just for their affordability but also for their technical prowess. At the same time, local populations are also very discerning of when to use of the allopathic system, as "patients consider the hospital as the only institution where emergency care can be provided and reliable diagnostic facilities are available" (Aïres et al., 2007, p. 571). This assessment of when to consult with these two approaches to wellness is similar to what we found in our initial study of the health sector in Northern Ghana (Gariba & Langdon, 2005) as well as echoing Ventevogel's (1996) observation, and runs contrary to the pejorative characterization of patients' choice-making ability in much of the other mainstream literature on either traditional medicine (c.f. Sarpong, 2008; Sittoe, 2008) or bonesetters (c.f. Onuminya, 2004; Van der Horst, 1985). Interestingly, the article also echoes our observation, introduced above, that there is a strong tradition of knowledge sharing and standardization of approaches amongst bonesetters:

> Our data reveals that treatments offered by different bonesetters in Ghana seem quite similar with regard to methods, materials and follow-up, and are comparable with methods described in the past.
>
> Aïres et al., 2007, p. 571

In this respect, Aïres et al. (2007) is an important addition of a more respectful dialogue between these two different approaches to healing broken bones, where "fracture treatment can serve as a model for respectful and efficient co-existence of traditional and biomedical medicine" (p. 564). This progressive stance leads Aïres et al. (2007) to conclude that it is not training but dialogue that is needed, something Onuminya and Sarpong put in jeopardy. However, despite this progressive stance, the study is still part of the same colonizing process: much as the allopathic method itself focuses largely on the ailment, rather than a holistic engagement with the lifestyle of the individual involved, this study largely focuses on the practice of bonesetters and on the opinion of patients rather than trying to see its root philosophy of wellness. In terms of power relations, even from the perspective of this last more sympathetic article, it is still through the lens of allopathic and biomedical epistemology that other epistemes of wellness, such as that of the bonesetters, are judged.

What is so fascinating in the logic presented by Arhin and Sarpong, as well as Onuminya in the Nigerian context, is the way 'traditional medicine' is recognized as an important contributor to the wellbeing of local populations, yet in the same breath, proponents of this logic call for "integrating" it "into the national health care system" (New Legon Observer, 2008, p. 1)—something we understand as a renewed attempt at colonization and containment. Indeed, we are arguing here that these types of discursive discussions of integration are a continuation of the coloniality of difference, where the mainstream allopathic system in Ghana is trying out new ways to discipline, co-opt, or contain these other paths to wellness because, as what Foucault (1980) would call "disqualified knowledges" (p. 82), their existence in the world undermines the truth claims of the allopathic way. After all, if either of the two systems—if they should even be dichotomized this way[2]—should adjust to change not only in its practice, but also in its whole philosophy of health and illness, it is the one that 70 percent of Ghanaians *don't* use. To pursue the reverse is to indicate the underlying political project at work: a global design attempting to subvert, coerce, and control an alternative set of local knowledges (Mignolo, 2000). This disciplining of differ-ence must be understood as part of a larger process of neoliberal globalization, where it serves the biomedical needs of pharmaceutical companies through the appropriation of Indigenous herbal knowledge—something Shiva (1997) has aptly called biopiracy.

Coming back to the remarks of the village elder mentioned in the opening, one way to destabilize this global design is to unpack how different perspectives shift the terrain of what is and is not traditional medicine. For instance, while Sarpong, Arhin, and Onuminya argue for integration as a regulatory need—due to charlatans and quacks—this elder sees commodification of service—seen both in hospitals and charlatan practice—as the greatest problem (Gariba & Langdon, 2005). For him, this commodification of well-being is the single biggest characteristic that separates the systems, and it leads him to conclude, "the two are not the same at all" (Agyeyomah & Langdon, 2009, p. 137). Sublimating these problems of commodity culture onto an alternative episteme of wellness—as well as ignoring the clearly wide-spread nature of this episteme and its longstanding formative process—is a convenient oversight that makes arguing for integrating bonesetting into allopathic training process all the easier.

THE PEDAGOGY OF THE PROVERB: FORMATION IN THE EPISTEMOLOGY OF BONESETTING

In our opening, we highlighted the need to move away from the false dichotomy implied by the language around the allopathic and traditional medicine divide. Calling the many people who contribute to wellness in Ghana from outside the allopathic fold TMPs, or in the case of bonesetters TBSs, ignores the way in which a wide variety of practices are lumped into this category called traditional. It is precisely this conflation of alternatives that allows the fraudulent practitioners out there to be labeled TMPs along with those deeply embedded in a practice with historical legitimacy and epistemic roots. It is in reaction to this conflation that we pointed out that there are important similarities in practice between bonesetters in multiple locations across the West African subcontinent. These similarities do not mean this approach can be misrepresented as the only alternative to allopathy, but it does mean that this truly is a system of approach to wellness based on shared philosophy, practices, and knowledge. In a contrast that reveals the alternative location from which this system of wellness emerges, Chief Isshaku Gumrana Mahamadu, the bonesetter at the centre of this study, uses proverbs with deep cultural relevance to describe the origin and learning process of his practice and its relationship with allopathy, instead of Westernized acronyms

Figure 16.1   Chief Isshaku Gumrana Mahamadu, assistants, and Loagri community members (photo: Coleman Agyeyomah)

that simplify rather than enrich. Before turning to unpack these proverbs we wish to first underscore the deep interconnection between this epistemology of wellness and its strong cultural roots by quoting here from our previous work with Chief Isshaku to describe his philosophy in detail:

> Chief Isshaku Gumrana Mahamadu is the leader of a small town called Loagri, just outside of Wale wale. His position as chief also makes him the head of his clan, from which his line of bonesetters come. . . . Chief Isshaku's clan has been practicing bonesetting in Loagri for six generations. The bonesetter operates in a web of other traditional wellness practitioners, including sooth-sayers who help diagnose the source of a problem, and herbalists who use generations of local environmental knowledge to create medications to help address ailments. These are not strict specialist boundaries, however, as in some cases—such as in Isshaku's clan—bonesetters serve as both herbalists and bonesetters. Their generations of knowledge of the local environment have afforded Loagri bonesetters a vast knowledge of different herbs that can be used to assist in the healing of broken bones. . . . [Despite changes in the environment and the kinds of bone fractures he sees] Chief Isshaku has not changed his underlying philosophy of patient care. Unlike the allopathic method, which aims to diagnose an ailment based on the logic of deduction derived from the symptoms at hand, Isshaku believes that treating a broken bone is only one element of patient care. The most important aspect of his treatment is connected to understanding the larger issues in a patient's life that led him or her to break his/her bone. This philosophic approach is demonstrated by the manner in which Isshaku describes the various conditions of the patients in his care. He will usually begin far before the event that led to them breaking a bone, and he may even connect it to issues embedded in the ancestral past, if he knows enough of the patient's history. The event that led to the break will be put in context, even as it becomes part of the patient's story of recovery as well. In this way, a broken bone is seen as only one episode in a full life (interview with Chief Isshaku, 2007).
>
> Agyeyomah & Langdon, 2009, pp. 142–143

We present this précis here to convey and emphasize both the origins and philosophy of Chief Isshaku's practice. In the section that follows we expand upon the specifics of the formation of bonesetters from Chief Isshaku's perspective. In doing so, we will focus on four main themes: (1) deepening information on origins of bonesetter profession and knowledge production; (2) the qualities a prospective bonesetter needs before being considered for training; (3) the stages in the formation process; and (4) issues concerning the potential integration of Isshaku's approach to wellness and the allopathic health care system. Before this though, it is important to briefly share our methodology used in researching these questions.

Over the past five years, all three of us have been engaging Chief Isshaku in discussions surrounding both his practice and his philosophy of wellness

(Coleman especially has taken the lead on this). Building on Randall's (1996) use of the concept of restorying, we have worked with Chief Isshaku to deepen our understanding of his craft through a process of reengaging with him on related themes a number of times over these five years. This process of reengagement, or restorying, allows meaning making to be deepened both through the process of returning to previous discussions for clarification, and through a reflective distancing that time between discussions can allow (Randall, 1996). For instance, while we discussed the origins of bonesetting in our previous engagement with Chief Isshaku (Agyeyomah & Langdon, 2009), here in our current round of restoried discussions, new insights emerged. Also, based on the long-term nature of our relationship with Chief Isshaku, we were able to begin our current round of discussions with him at a much more detailed level than was previously possible, especially concerning the formation processes of bonesetters. As Aïres et al. (2007) noted, there is a great deal of mistrust of researchers by those practising a wellness approach other than the allopathic method. Thus, we take this relationship of trust seriously, unlike Onuminya (2004). The new round of discussions conducted over several dates in March 2009 had as its goal a restorying of the description of bonesetter formation sketched out in our last set of dialogues (captured in Agyeyomah & Langdon, 2009). Each of the four guiding themes is explored below through a proverb Chief Isshaku used in restorying them, except the last proverb, which is drawn from the title of this chapter and comes from the elder mentioned above.

A crab cannot deliver a bird—(origins of bonesetting)

In deepening the discussions around the origin of the bonesetting profession, Chief Isshaku connected it to an evolving traditional Mamprugu clan structure where professions were associated with particular clans. He cited the examples of butchers and blacksmiths as two other clan-professions, like the herbalist/ bonesetter clan. There is clear knowledge reproduction logic to this, according to Isshaku, as it ensures knowledge transfer to the next generation. This knowledge has been traditionally passed on through three methods: oral history, training/coaching, and mentoring. In a clear sense, the specialization associated with this process of clan-based learning ensured there would always be these skills and this knowledge available to the community. Isshaku also emphasized how this structure meant continuity of the quality of the profession. Members of other clans would be at a disadvantage because they would not be exposed to this profession. He captured this logic of specialization in the above proverb, "a crab cannot deliver a bird" (interview, Chief Isshaku, 2009).

The fly that fails to heed to advice follows the corpse to the grave—(criteria to become bonesetter)

Despite this history of clan specialization, Chief Isshaku underscored that being a member of the clan does not mean one can automatically become an apprentice to the bonesetter. There is a set of criteria (described below) that a person has to meet before they can be considered. More recently, he noted, some flexibility has developed in allowing those who meet all the criteria yet are not of the clan can also be taken on as apprentices—a change that underscores the evolving nature of this knowledge production and formation system—though the profession is still male only. The end goal of the process of evaluation is to ensure that a potential candidate has the right character to be able to take the bonesetters spiritual oath (discussed below) and live by it. While Chief Isshaku admits this criteria evaluation system is not foolproof and that there are some who break the oath and charge people for their services, these oath breakers also know they are playing with fire—a sentiment he captured in the proverb above and which is underscored in the formation process that apprentices go through even before they are allowed to take the oath.

Chief Isshaku described each of the key criteria in the following way:

- *Honesty and truthfulness:* required because a bonesetter must be honest with his patients and must never charge for his work (a prescription that comes from his forebears). Charging for healing will lead to loss of potency of the treatments a bonesetter uses. Additionally, ensuring a candidate is honest also forestalls any chance the future bonesetter will dilute his herb ointments, or defraud a patient through a false prescription.
- *Hardworking and perseverant:* attitude is crucial to being a bonesetter as well, where the interests of those who seek help always come before personal interests. This attitude is also crucial in today's climate conditions, where climate change has forced bonesetters to go further a field to find the herbs they require for their practice.
- *Patience, tolerance, and sympathy:* required in dealing with those who seek help from a bonesetter. All manner of people come to the bonesetter and Chief Isshaku noted one needs a "heart of a lion" to be able to work them towards healing. Sympathy is especially critical as it helps a bonesetter see what has befallen a patient as one chapter in a history of their whole lives. Consequently, Chief Isshaku noted that while pain is an important part of the healing process in his practice, he has also begun working with allopathic practitioners on providing pain relief for some patients. Along with this, he underscored that the practice requires bonesetters to identify with those who seek help, and that this help is more than just mending bones, but is about addressing the mindset that may have led to the broken bone in the first place.
- *Aptitude for learning:* one has to demonstrate a total commitment to learning; additionally one must have a quick retentive memory and be able to differentiate the different types of herbs and their uses in the healing process. This forestalls a situation where one could misapply the herbs thereby compounding the problem of the patient.

- *Humility*: Chief Isshaku indicated the overriding quality for being a bonesetter is humility. One needs to be humble in his dealings with the patient. There is no room for persons who are boastful arrogant and disrespectful to their patient. Bonesetters need to treat every patient equally and not discriminate. He indicated that not charging for service is an important way to forestall discriminative tendencies. (Interview with Chief Isshaku, 2009)

Without patience one cannot dissect an ant let alone diagnose what is within the entrails—(Bonesetting formation process)

In our previous discussions with Chief Isshaku, he told us that it could take more than twenty years to become a full-fledged bonesetter (Agyeyomah & Langdon, 2009). In our current discussions, we returned to this number and asked him to break down the stages of this process. He identified five steps in the apprenticeship stage, and then a further stage of formal bonesetting training. By his account, the apprentice stage can take a decade and a half or more, depending on the apprentice, and the formal training can take another three to four years. All of these numbers are estimates because some apprentices never graduate to the formal stage. The five stages of apprenticeship are:

- *Initiation stage:* For two to three years after candidates within the clan indicate their interest in becoming bonesetters, and after they have met all the criteria, the candidates go through an initiation. They are still under observation to see if they will make good apprentices. During this period they first observe what the bonesetter does, and then are tasked with more menial aspects of patient care (cleaning patients sores, helping them go to the bathroom, etc.)
- *Post initiation stage:* After they have passed through the initiation, apprentices are introduced to the various herbs used by the bonesetter and their uses. After this, they are tasked with collecting these herbs. Once they are conversant with all of this, they are taught how to prepare the herbs for treatment.
- *Hands on training of Bonesetting:* Apprentices then begin to learn the different fractures. It is quite awhile before they are permitted to deal with simple fractures and the application of herbs, and even then they are under strict supervision.
- *Evaluation stage:* Apprentices are then evaluated, not only for their competence in these tasks but also for the way in which they continue to reflect the core criteria of a bonesetter. Even if they have excelled in the skills, the apprentice may be dismissed for not meeting even just one of the criteria discussed above.
- *Spiritual Oath:* After this evaluation apprentices begin formal bonesetter training. Yet, before this happens they must swear an oath before the

master bonesetter, as well as the fetish priest and local chief to always place the health of others before one's own self-interest. This oath is also sworn before the whole community, so that any who break the oath are jeopardizing not only their relationship with their master, but with the whole community. In connection with this, Chief Isshaku remarked, "above all, news will travel far and wide for people not to deal with [an oath breaker] as a herbalist and bonesetter" (interview with Chief Isshaku, 2009).

After the oath is taken, formal training begins. It involves a more detailed formation in herb gathering, preparation, and storage, bonesetting, positioning the factures, and the application of dressing and medication. This stage takes three to four years before the apprentice is deemed a qualified bonesetter. Yet key to this entire process is a reemphasis of the above criteria; completion is only possible when a master bonesetter is completely satisfied that the apprentice is ready. It was in the context of the patience needed to go through this process that Chief Isshaku gave us the above proverb, "Without patience one cannot dissect an ant let alone diagnose what is within the entrails" (interview with Chief Isshaku, 2009).

To die is honey, to live is salt—(allopathic integration)

Unlike the elder who gave us the title proverb of this chapter, Chief Isshaku sees some potential for working with the allopathic system. Key, however, for him is that this other system of health care recognizes the different philosophy behind his practice. This goes beyond issues of paying for service, as any one who comes to the bonesetter is immediately immersed in a whole approach to wellness where payment is secondary. For instance a patient must also consult a soothsayer to place their current ailment in the context of their lives, and even the lives of their forebears. Guidance is also sought from ancestors as well as gods on treatment paths. In this sense, healing is physical, mental, and spiritual. Nonetheless, Chief Isshaku believes that if allopathic practitioners could accept and respect these other dimensions to the practice there could be mutual collaboration. The bonesetting practice could even be situated in an associated institution. Yet, he was adamant that it could never be under the allopathic system because this could jeopardize the independence of his practice. He also raised big concerns about the potential to institutionalize the formation of bonesetters in an allopathic institution. To him, this is merely another form of allopathy and an elimination of the independence of his practice. This critique notwithstanding, he did see some areas where the allopathic system could be of immediate help to his work, especially in the area of anaesthesia. In pursuing a path directed at making his patients' lives better, he sees pain relief as a major benefit to their experience at the bonesetter. Yet the bottom line for Chief Isshaku is respect. There can be no healing without respect—something the elder's proverb "to die is honey, and to live is salt" implies. If one must degrade her/himself to be healed, then really there has been no healing. Likewise, if bonesetting practice in Ghana must become subservient to the logic of allopathy in order to end its supposed marginalization, then there is no end of marginalization at all. Chief Isshaku can see this; the elder and other users of this system can see it; and yet allopathic practitioners and researchers cannot.

CONCLUSION

Despite overwhelming numbers that suggest Ghanaians value other approaches to wellness over that of the national allopathic healthcare system, healthcare researchers, practitioners, educators, and administrators believe it is these other systems of wellness that must adjust and be integrated into the national healthcare envelope. Similarly, despite announcements celebrating the end of the subjugation of these alternative ways of promoting wellness by many of these same allopathic researchers and educators, the same forms of subjugation of knowledge and coloniality of difference still persist. This is especially true in discussions of training, where university administrators are making all-out bids to submerge these alternative sources of wellness, as well their learning processes, in their Western dominated fold. In this article, we have presented a contrasting alternative formation process that is embedded in a philosophy of wellness that places the character of a practitioner before anything else—helping to ensure patients will always be treated with dignity and respect. With this contrast in mind, we ask again, which system should be learning from which?

Acknowledgment Note: We all wish to acknowledge and deeply thank Chief Isshaku Gumrana, who created the space and opportunity for us to undertake the research contained in the pages that follow.

REFERENCES

Anderson, B. (1991). *Imagined communities: reflections on the origin and spread of nationalism.* London: Verso.

Arhin, P. (2008). Policy for integrating natural products into National Health System. *New Legon Observer, 2*(15), 4–6.

Ariës, M. J. H., H. Joosten, H. H. J. Wegdam, & S. van der Geest (2007). Fracture treatment by bonesetters in central Ghana: Patients explain their choices and experiences. *Tropical Medicine and International Health, 12*(4), 564–574.

Carothers, T. (2002). The end of the transition paradigm. *The Journal of Democracy, 13*(1), 5–21.

Eshete M. (2005) The prevention of traditional bonesetter's gangrene. *Journal of Bone and Joint Surgery—British Volume, 87-B*(1), 102–103.

Foucault, M. (1980). *Power/Knowledge: Selected interviews & other writing.* New York: Pantheon Books.

Hoff, W. (1992). Traditional healers and community health. *World Health Forum, 13,* 182–187.

McMichael, P. (2004). *Development and social change: A global perspective.* London: Sage.

Mignolo, W. (2000). *Local histories/global designs: Coloniality, subaltern knowledges and border thinking.* New Jersey, NJ: Princeton University Press.

New Legon Observer (2008). Why are so many people using herbal medicine and what are the implications. *New Legon Observer, 2*(15), 1–3.

Onuminya, J. (2004). The role of the traditional bonesetter in primary fracture care in Nigeria. *South African Medical Journal, 94*(8), 652–658.

Onuminya, J. (2006). Performance of a trained traditional bonesetter in fracture care treatment. *South African Medical Journal, 96*(4), 315–322.

Randall, W. (1996). Restorying a life: Adult education and transformative learning. In J. E. Birren (ed.), *Aging and biography: Explorations in adult development* (pp. 224–247). New York: Springer.

Sarpong, K. (2008). Tertiary level education and training for herbal medicine practitioners. *New Legon Observer, 2*(15), 15–18.

Shah, R. K., V. K. Thapa, D. H. A. Jones, & R. Owen (2003). Improving primary orthopaedic and trauma care in Nepal. *Education for Health, 16*(3), 348–356.

Shiva, V. (1997). *Biopiracy: The plunder of nature and knowledge.* Cambridge, MA: South End Press.

Sittie, A. (2008). Renaissance of herbal medicine and safety concerns. *New Legon Observer, 2*(12), 32–36.

Ventevogel, P. (1996). *Whiteman's things: Training and detraining healers in Ghana.* Amsterdam: Het Spinhuis.

Warren, D. M. (1974). Bono traditional healers. *Rural Africana, 26,* 25–34.

Warren, D. M., G. S. Bova, M.A. Tregoning, & M. Kliewer (1982). Ghanaian national policy toward indigenous healers. The case of the primary health training for indigenous healers (PRHETIH) program. *Social Science and Medicine, 16,* 1873–1881.

NOTES

CHAPTER 3

1. Bolivia has actually adopted the declaration as national law, which is exceptional.
2. The reversal of this is of course "primitivism," which reproduces this temporality often with even greater efficiency.
3. In Jharkhand, we have the early introduction of "customary land laws" based governmentality in the form of the Southwest Frontier Agency under the Wilkinson Rules in 1834 (Jha, 1987).
4. Also see the volumes of *Subaltern Studies,* especially in the 1980s.
5. I have particularly focused on the story of the politics of indigeneity in the Jharkhand region of India. Not only has Jharkhand led the Indian participation in the transnational indigenous movement but also in the entire colonial period it was the region with the most pronounced development of governmental technologies aimed at indigenous populations in India and the most active history of indigenous political agency. This had led to it leaving a most decided imprint on the Indian legal, scholarly, and activist imagination of adivasi indigeneity, including customary law regimes, the formation of the discipline of anthropology in India, and the volume of scholarly attention devoted to the indigenous question.
6. See Mahato (1971), Roy (1995), and De Sa (1975) for introductions to the question of conversion and adivasi political histories and consciousness. The definitive volume on adivasi Christianity is still awaited.
7. Of course the loopholes in such policies are many and they keep the possibility of displacement the central political theme in adivasi (and now in general Indian rural) life.
8. I place the current series of confrontations between the Maoists and the state in India within this larger complexity of indigeneity or adivasiness. Obviously, newer elements of indigenous politics are also emerging in this Maoist moment, but it is deeply connected to these histories of adivasi modernity and politics. In fact, I think the present "Maoist" moment can be better grasped through developing a sense of an adivasi idiom of modernity as opposed to searching in the formal historiography of Maoism in India.
9. Also see Karlsson (2003) and Ghosh (2006) for different discussions of this.

CHAPTER 4

1. It would be a gross misrepresentation to characterize all Indigenous Peoples' positions as being inherently anticapitalist and anticolonial. There are many internal debates and conflicts among Indigenous Peoples regarding values and models of "development" (see, for example, Bargh, 2007; Kelsey, 1995, 1999 for more on this in Aotearoa/New Zealand). Some, particularly those engaged in business enterprises, embrace free market/free trade policies as opportunities to create business and economic relationships locally and internationally, sometimes arguing that economic globalization allows them to operate somewhat independently of the colonial nation-state. Here, however, I refer to those movements, communities, networks, and mobilizations that critique and oppose neoliberalism and frame this in the context of anticolonial struggles for self-determination.

2. Aotearoa is the Maori word for New Zealand. I use Aotearoa/New Zealand when referring to the country, and New Zealand when referring to the government or official institutions.

3. The meeting was co-organized by GATT Watchdog, an activist/research group for which I was an organizer for many years.

CHAPTER 5

1. E.g., the Kaptai Hydroelectric Power project.

2. I express my sincere thanks to Mr. Golam Iftikhar Hussain, Graduate Teaching Assistant, for his help while conducting the research.

3. A livelihood framework, retrieved from, http://www.livelihood.wur.nl/?s=D1-Alivelihhodframework

4. When the people used the term *Magh*, they usually identify it with pirates and anarchists. See Dewan (1990) and Khan (1999).

5. This paper mill was constructed in Chandraghona at the cost of approximately US$13 million, including US$ 4.2 million from the World Bank. See Gain (2000).

6. The dam was constructed in 1964 with USAID (United States Agency for International Development) funds. It submerged 250 square miles of farming land in the hilly districts. See Gain (1998).

7. The institution did not consult its target people while implementing its projects. See Tripura (2000).

8. Personal observation, 2004.

9. Ibid.

CHAPTER 6

1. In this chapter, Formosa refers to the island and its original inhabitants. Taiwan refers to the subsequent settler society and its state-centric

political development. There is no word for the island in the Truku language, which developed and matured in the central mountains far from the sea.

2. Eventually, I was taken to visit an elderly shaman. She gave me the name Walis Watan, saying that I have a similar personality to someone she knew with that name during the Japanese period.

3. Retrieved from, http://www.apc.gov.tw/main/docDetail/detail_official. jsp?isSearch=&docid=PA000000001795&linkSelf=231&linkRoot=231 &linkParent=231&url=

CHAPTER 9

1. According to the Swayambhu Purana, the Kathmandu Valley was a giant lake called Nāgdaha until the Bodhisattva Manjushree cut open part of the southern hill of Kachchhapāla with the aid of a holy sword called *Chandrahrāsa,* and then cut open Gokarna daha (lake) and drained the giant lake, allowing Newars to settle the valley land.

2. This refers to the introduction of Sanskrit vocabulary in another language or dialect.

3. Nepal Bhasa (also known as Newa Bhaye and Newari) is one of the major languages of Nepal. It is one of roughly 500 Sino-Tibetan languages and belongs to the Tibeto-Burman branch of this family. It is the only Tibeto-Burman language to be written in the Devanāgarī script.

4. In the late fifth century, rulers calling themselves Licchavis began to record details on politics, society, and economy in Nepal.

5. King Jaya Sthithi Malla of the Malla dynasty reined the Nepal Valley from 1372 to 1395 and introduced numerous social reforms including the first major codification of caste laws.

6. A benevolent community trust based on caste or kinship links whose basic function is to look after and maintain temples and fountains, organize festivals, and take care of cremations.

CHAPTER 10

1. The only-in-English proverbs cited are mentioned in Ruth Finnegan's *Oral Literature in Africa* (1970), in the chapter on proverbs (pp. 389–425). The rest (i.e., those translated from the Somali) are ones I am personally familiar with from the Somali culture, and I am literally or figuratively translating them.

CHAPTER 11

1. CMT, DDS: Pastapur.

CHAPTER 12

1. The Krishna-Bhakti Movement spread to southern India by the ninth century AD, while in northern India Krishnaism schools were well established by eleventh century AD.
2. *Tariqah* means way, path, or method and refers to an Islamic religious order; in Sufism, it is conceptually related to *haqiqah* (truth), the ineffable ideal that is the pursuit of the tradition.
3. Zen emphasizes experiential Pranja—particularly as realized in the form of meditation known as *zazen*—in the attainment of awakening, often simply called the path of enlightenment.
4. *Fakir* is the term often used in English to refer to Hindu ascetics (e.g., sadhus, gurus, swamis, and yogis) as well as Sufi mystics. It can also be used pejoratively, to refer to a common street beggar.
5. The only evidence is an obituary note in a local newspaper in October 1890.
6. *Sadhana* (Sanskrit) is a term for "a means of accomplishing something" or more specifically "spiritual practice."
7. In Hindu traditions, *Sadhu* is a common term for an ascetic or practitioner of yoga (yogi) who has achieved the first three Hindu goals of life: *kama* (enjoyment), *artha* (practical objectives), and even *dharma* (duty).
8. *Sufi* is generally understood to be the inner, mystical dimension of Islam. Another name used for the Sufi is seeker.

CHAPTER 16

1. Ventevogel (1996) describes a program for *Primary Healthcare Training of Indigenous Healers* (PRHTIH), implemented in Techiman, Ghana, in 1979. This program was of limited success, due to a profound lack of trust between TMPs and Westernized systems. Ventevogel spoke with a number of TMPs involved in the study who perceived this initiative as an attempt at "brainwashing." Most prevalent, however, was a general mistrust of the allopathic system, as TMPs wondered how these government people could possibly presume to tell them about methods that have been in their collective consciousness for hundreds of years.
2. As Mignolo (2000) puts it, "If Western Cosmology is the historically unavoidable reference point, the multiple confrontations of two kinds of histories [local ones and ones purporting to be global designs] defy dichotomies" (p. ix).

LIST OF CONTRIBUTORS

Ali A. Abdi is Professor in the Department of Educational Policy Studies, University of Alberta, Canada. His areas of research include comparative and international education; citizenship and development education; cultural studies in education; African philosophies of education; and postcolonial studies in education. He has published in journals such as *Comparative Education, Compare, McGill Journal of Education, Journal of Black Studies, Journal of Educational Thought, International Education* and *Journal of Postcolonial Education*. He is the author of several books and coedited collections including *Education and Development in South Africa* (2002), *Politics of Difference: Canadian Perspectives* (2004), *Issues in African Education: Sociological Perspectives* (2005), *African Education and Globalization: Critical Perspectives* (2006), *Educating for Human Rights and Global Citizenship* (2008), and *Global Perspectives on Adult Education* (2009).

Coleman Agyeyomah is the Executive Director of Venceramos Consulting, with over twenty years of experience in community and local development. He also recently completed a Masters degree at Leeds University in the UK, where he critically examined conventional notions of local development, especially from the perspective of indigenous epistemologies. Although this masters was an important opportunity to reflect on the current state of the development industry, from his perspective, the only education that has ever mattered has been what he garnered through years of working with communities throughout Northern Ghana.

Bijoy P. Barua (PhD, University of Toronto) is Associate Professor and Chair in the Department of Social Sciences at East West University, Dhaka, Bangladesh. He offers courses at graduate and undergraduate levels in Development Studies, Research Methodology, Civil Society, Ecology, Indigenous Knowledge, Development Management, Gender and Development, and Sociology. He has also worked as researcher, teacher, trainer, program manager, and consultant in Bangladesh, India, Nepal, Sri Lanka, Thailand, Vietnam, Ghana, Switzerland, and Canada. His research interests include international development; comparative education; participatory research; ethnic minorities; indigenous knowledge; engaged Buddhism/ecology; community health; gender and culture, development and civil society. He has published in academic journals such as *International Education* (USA), the *Canadian Journal of Development Studies, Medicus Mundi* (Switzerland), *Managerie* (Germany), and *the Parkhurst Exchange* (Canadian Medical Educational Journal for Doctors). He has also contributed to the edited collections *Global Perspectives on Adult Education* (2009), *Education Decolonization and Development* (2009), and *Education, Participatory Action Research and Social Change: International Perspectives* (2009).

Rebecca Butler participated in the Trent in Ghana program in 2001 where she had the opportunity to work with Bonesetter Chief Isshaku Mahamadu and his community of Loagri. Using participatory techniques, she helped to organize the construction of a small building to house patients and their families during treatment. This project gave her the opportunity to talk in detail with Chief Mahamadu about bonesetting as a crucial part of health care in Ghana. Rebecca has recently completed her Master of Planning focusing on small-scale community development techniques. She currently works as a researcher and writer in Toronto, Canada.

Aziz Choudry is Assistant Professor in International Education, Department of Integrated Studies in Education, McGill University, Montreal, Canada. He is author (with Jill Hanley, Steve Jordan, Eric Shragge, and Martha Stiegman) of *Fight Back: Workplace Justice for Immigrant Workers* (Halifax: Fernwood, 2009), and is coeditor (with Dip Kapoor) of *Learning from the Ground Up: Global Perspectives on Social Movements and Knowledge Production* (New York: Palgrave Macmillan, forthcoming, 2010). A longtime anticolonial, social, and environmental justice activist, writer, and researcher, he has worked extensively in numerous NGOs, movements, and activist groups in Aotearoa/New Zealand, throughout the Asia-Pacific, and Canada. His articles have appeared in, among other publications, *Canadian Journal for the Study of Adult Education, International Education, McGill Journal of Education*, and *Cultural Studies of Science Education*.

Kaushik Ghosh teaches Anthropology and Asian Studies at the University of Texas at Austin. He was born in Calcutta and did his graduate work in anthropology at Stanford and Princeton Universities. He has been deeply involved with adivasi indigenous movements against displacement in India, especially in the state of Jharkhand. Based on this work, he is currently finishing a book, *Indigenous Incitements: Political Form and Adivasi Struggles against Dispossession in India.* He also writes on theories of governmentality, the history of adivasi labor migration, adivasi religion, and the culture of big dams. Some of his key publications have appeared in *Subaltern Studies* and *Cultural Anthropology.* His recent work has turned toward a theorization of the frantic speculative global economy looked at from the condition of relentless dispossession of both adivasi and other rural populations in India. An academic nomad, he splits his time between Austin, Delhi, Jharkhand, and Calcutta.

Ehsanul Haque is an Associate Professor in the Department of International Relations, University of Dhaka, Bangladesh. He also teaches in the Department of Social Sciences, East West University as an adjunct faculty. He obtained an MA in International Affairs from Ohio University, with a specialization in Southeast Asian Affairs. He has published on issues like South Asian peace process, Chinese military power, democracy movement in Burma, ASEAN and regional security, North Korean nuclear program, ethnicity in South Asia, and conflict and poverty. From 1994 to 2003 he was the Associate Editor of *Journal of International Relations* (a biannual journal of the Department of International Relations, University of Dhaka) and from 1995 to 1998 he served as the Assistant Editor of *Social Science Review* (a biannual journal of the Faculty of Social Sciences, University of Dhaka). He is a recipient of the US Department of State Fulbright Scholarship.

Rayyan Hassan is currently a Senior Lecturer at the Department of Social Science, East West University, Dhaka, Bangladesh. He obtained his masters degree in Social Change and Development from the University Of Wollongong, New South Wales, Australia. He has written various research reports and journal articles. His research interests include environmental education, and indigenous knowledge of primary resources, culture, and sustainable development. He has worked in both multilateral international agencies and national organizations in the development sector.

Dip Kapoor is Associate Professor, International Education, Department of Educational Policy Studies, University of Alberta, Canada and Research Associate, Center for Research and Development Solidarity (CRDS), an Adivasi-Dalit people's organization in India. His areas of research include political-sociology of education and development; globalization, colonialism and education; indigenous social movements, NGOs and popular education/learning; global education; and participatory/qualitative research methods. He is editor/coeditor of the following book collections *Education, Development and Decolonization: Perspectives from Asia, Africa and the Americas* (2009), *Global Perspectives on Adult Education* (2009), *Education, Participatory Action Research and Social Change: International Perspectives* (2009), *Learning from the Ground Up: Global Perspectives on Social Movements and Knowledge Production* (in press), and *Beyond Development and Globalization: Social Movement and Critical Perspectives* (forthcoming).

Jonathan Langdon has been engaged in working with local organizations and movements in Ghana for the last ten years. This work has always been grounded in privileging local notions of development, learning, and knowledge over those determined by outside agencies—be they at the national or international level. In the Ghanaian context, the majority of his work has been based in Northern Ghana—the most underprivileged section of the country. This work has consisted of, amongst others, an assessment of health, education, and local government services from community perspectives, and a study of the ways in which indigenous forms of justice are accessed by the poor. He now lives in Antigonish, Nova Scotia, where he is an Assistant Professor of Development Studies at St. Francis Xavier University.

Christine Hellen Mhina obtained her PhD in International Cultural Studies at the University of Alberta. Over the years of her work experience as Agricultural Extension Officer she has learned the value of including diverse perspectives in action for change. As part of her dissertation work, Christine engaged in a participatory action research with a local community in Tanzania with a focus on indigenous women's knowledge, agency, and social change. She has published book chapters that appear in *Global Perspectives on Adult Education* (2009), *Education, Participatory Action Research, and Social Change* (2009), and articles in *Canadian Journal of University Continuing Education* and *Higher Education Perspectives*. Her teaching and research interests are in participatory democracy; women's emancipation; ethnic minorities; and indigenous knowledge. Christine teaches *Swahili language* at the University of Alberta and *Aboriginal Societies* at Concordia University College of Alberta. She is also the Director of Diversity program at Sexual Assault Centre of Edmonton.

Sourayan Mookerjea's research addresses contradictions of globalization, migration, urbanization, subaltern social movements, popular culture, and class politics. Recent publications include *Canadian Cultural Studies: A Reader,* coedited with Dr. I. Szeman and G. Faurschou, (Duke University Press, 2009). Dr. Mookerjea is Associate Professor of Sociology at the University of Alberta, Canada.

Sudhangshu Sekhar Roy is Assistant Professor in the Department of Mass Communication and Journalism at the University of Dhaka. He also teaches in the Department of Social Sciences of East West University as an Adjunct Faculty. He has written several books and contributed to various articles to various national and international journals. He is the fellow of International Institute for Journalists (IIJ) of Berlin, Germany, and is the president of IIJ, Bangladesh Chapter. His research interests include indigenous media and cultural heritage; freedom of press, speech, and expression; political and media systems; and media ethics.

Deepa Shakya obtained her Masters in Development Studies from East West University, Bangladesh and was conferred with a gold medal for her academic performance. Her thesis was titled *Universal Education, Culture and Socio-Economic Development in Kathmandu Valley: A Case of Newars in Nepal.* She has pursued Masters in Business Studies from Tribhuvan University, Nepal, and is currently working on her thesis titled *Nepal's Export Trade and Export Marketing: Special Reference to Bangladesh.* At present she is working as Institutional Processes Analyst, providing management consultancy to different business houses in Nepal. She is also involved as part time faculty member in National College Center of Development Studies and Ace Institute of Management, Nepal. She has worked as freelance journalist, project coordinator, and study abroad counselor in Nepal. Her research interests lie in the area of sustainable development, education, and indigenous culture and knowledge. She speaks fluent English, Nepali, Newari, Hindi, and Bengali.

Edward Shizha is an Assistant Professor in Contemporary Studies and Children's Education and Development at Wilfrid Laurier University (Brantford) in Canada. His academic interests are in contemporary social problems and education including globalization, postcolonialism, and indigenous knowledges in Africa. He has published refereed articles that appeared in international journals such as *International Education Journal, Alberta Journal of Educational Research, AlterNative: An International Journal for Indigenous Scholarship*, and *Australian Journal of Indigenous Education* and book chapters published in *Issues in African Education: Sociological Perspectives* (Palgrave Macmillan, 2005), *African Education and Globalization: Critical Perspectives* (Lexington, 2006), *Global Perspectives on Adult Education* (Palgrave, 2008), *Education and Social Development: Global Issues and Analysis* (Sense, 2008), and *International Perspectives on Education, PAR and Social Change* (Palgrave Macmillan, 2009). He has coauthored *Citizenship Educational and Social Development in Zambia* (Information Age Publishing, 2010), and *Educational Development in Zimbabwe: The Social, Political and Economic Analysis* (Sense, 2010).

Scott Simon is Associate Professor in the Department of Sociology and Anthropology, University of Ottawa, Canada. He is author of two books about Taiwan: *Sweet and Sour: Life-worlds of Taipei Women Entrepreneurs* (Rowman & Littlefield, 2003) and *Tanners of Taiwan: Life Strategies and National Culture* (Westview, 2005). He has researched Taiwan since 1996, when he conducted PhD research for two years in Tainan. He has also done research in Taipei. With the support of the Social Science and Humanities Research Council of Canada, he conducted eighteen months of field research from 2004 to 2007 with the Truku of Hualien County and the Seediq of Nantou County, Taiwan. In 2008, he was a visiting scholar at the *Institut d'Asie Orientale* at the *École normale supérieure* in Lyon, France, where he wrote an ethnographic manuscript on indigenous-state relations in Taiwan. He is currently President of the Canadian Asian Studies Association.

Lantana Usman is currently a tenured Associate Professor at the School of Education, University of Northern British Columbia, Prince George, Canada. She graduated with a PhD in Educational Administration, Leadership, and Policy studies from the University of Alberta, Edmonton, Canada. She has published in peer reviewed journals and also has several book chapters related to her research and teaching area of educational policy studies, gender and education, children and youth schooling, international comparative and developmental education, economics of education multicultural education, social studies, and immigration education.

Njoki Nathani Wane (PhD, University of Toronto) is Associate Professor, current Director for the *Centre for Integrative Anti-Racist Research Studies*, as well as Director of Teaching. Her teaching areas and research interests include indigenous knowledges, anticolonial thought, antiracist education, spirituality, African Canadian feminisms, and ethnomedicine. Her most recent selected works include a coedited collection, *Theorizing Empowerment: Perspectives from a Canadian Feminist Thought* (Massaquoi & Wane, 2007) as well as chapters in *Integrating Traditional Healing Practices into Counseling and Psychotherapy* (Moodely & West, 2005) and *Anti-Colonialism and Education: The Politics of Resistance* (Dei & Kempf, 2006). She has also published in journals such as the *Journal of Race, Ethnicity and Education*.

INDEX

Printed in the United States
By Bookmasters